D1084984

NEWS FROM MARS

Science and Culture in the Nineteenth Century

BERNARD LIGHTMAN, EDITOR

NEWS FROM MARS

Mass Media and the Forging of a New Astronomy, 1860–1910

Joshua Nall

UNIVERSITY OF PITTSBURGH PRESS

Published by the University of Pittsburgh Press, Pittsburgh, Pa., 15260
Copyright © 2019, University of Pittsburgh Press
All rights reserved
Manufactured in the United States of America
Printed on acid-free paper
10 9 8 7 6 5 4 3 2 1

Library of Congress Cataloging-in-Publication Data
Names: Nall, Joshua, 1982– author.
Title: *News from Mars: Mass Media and the Forging of a New Astronomy,*
    1860–1910 / Joshua Nall.
Description: Pittsburgh, Pa.: University of Pittsburgh Press, [2019] |
    Series: Science and Culture in the Nineteenth Century | Includes
    bibliographical references and index.
Identifiers: LCCN 2019020482 | ISBN 9780822945529 (hardcover)
Subjects: LCSH: Mars (Planet) —Popular works. | Martians in mass
    media—Popular works. | Life on other planets—Popular works. |
    Astronomy—History—19th century—Popular works. |
    Astronomy—History—20th century—Popular works.
Classification: LCC QB641 .N275 2019 | DDC 523.43—dc23
LC record available at https://lccn.loc.gov/2019020482

Cover art: *detail*, 8-inch globe of Mars, by Ingeborg Brun after the maps of
Percival Lowell, Svendborg, Denmark, 1913. Courtesy the Whipple Museum
of the History of Science.
Cover design: Joel W. Coggins

To my brother, mother, and father

CONTENTS

NEWS FROM MARS

MARS

after Lowells Planiglob
(ca 1905)

ACUBE

MARE
CIMMERIUM
HESPERIA

ABAS

FAURUS

HELISSON

CAESUS

LAERON

CYCLOPUM
LUCUS

SINUS
GOMER

SURL

SRUS

CYCLOPS

CERBERUS

LAESTRYGON

TARTAROS

LUC
LUCRINUS

POLYPHEMUS

AETHIOPS

ORCHUS

PRIPUS
LUCUS

VIA ETERNA

MYRLAEUS

TRIVIUM
CHARONTIS

FONS
IMMORTALIS

INTRODUCTION

In the autumn of 1909, as the planet Mars swung away from its closest approach to Earth in seventeen years, two astronomers got into a fight. It was over the control of news from Mars. Both men were senior figures in their field, and both wanted the exact same thing: to secure for their own observatories exclusive control over the distribution of telegraphic bulletins reporting breaking astronomical news. With victory in this battle came the responsibility to collect observers' reports, adjudicate what counted as newsworthy, and then broadcast this selected information around the world, transmitting it to all subscribing observatories and news agencies, sometimes within hours of its receipt. Since 1882 Harvard College Observatory had performed this task in North America, partnering with the Royal Observatory in Kiel, Germany, to collect and rapidly redistribute announcements of planetary, asteroid, satellite, and nova discoveries.[1] But after a quarter century of uninterrupted broadcasts, Harvard College Observatory's director, Edward Charles Pickering, found himself and his institution facing down a coup. Two and a half thousand miles away, high on the desolate Colorado Plateau, Percival Lowell plotted a takeover.

At the heart of this battle was a disagreement over a single fundamental question: was there evidence of life on Mars? Lowell thought that the answer was yes, and in 1894 he had built a state-of-the-art observatory in Flagstaff, Arizona, with the express intention of proving it. Pickering disagreed, and his frustration with Lowell's bold claims about intelligent Martian life was soon exacerbated by their growing popular appeal. Lowell's success, Pickering realized, was predicated in part on his rival's access to Harvard Observatory's bulletin service, which simultaneously linked the Flagstaff astronomer to his colleagues and to the world's mass media. So Pickering wrote to Lowell to inform him of a change of service. Harvard Observatory would no longer redistribute any reports relating to Mars sent in to Cambridge from Flagstaff, ostensibly on the grounds that they were not time sensitive. But as Lowell knew very well, the rare close approach of Mars in 1909 put the planet in a prime position for making fresh observations and perhaps even revealing new details about its inhabitants. Reports of this work would very much be news, and their value as such would depend upon their ability to fit the time-critical demands of a telegraphically networked press. As far as Lowell was concerned, there could be no question that Pickering's move was anything other than a deliberate act of media sabotage, which could only be countered in kind. So in a cabled notice sent simultaneously to the press and to his American colleagues, Lowell "issued notice . . . that by an arrangement with the Centrale Stelle at Keil, Germany, the Lowell Observatory instead of Harvard hereafter will be the distributing center for planetary news in America."[2] Pressed by local journalists for a response, Pickering simply stated that no such takeover had occurred and that Harvard Observatory would continue to provide its own bulletin service as usual. In a bullish interview with the *Boston Journal*, the Harvard director openly conceded that this was a fight about news from Mars: "I presume the cause of the [takeover bid] can be found in my refusal to accept or reject Professor Lowell's claims relating to the habitation of Mars." In this matter, Pickering retorted, Lowell "is as safe from denial of his claim as is the man who has offered $1,000 for proof that under the old Charles River Park is the center of the world."[3]

This contest was defined by two intertwined problems. One was about Mars, and the other was about media. This book tells the story of this entanglement, why it happened, and what it meant for the practice of astronomy in that era. It explores the birth of two closely related new disciplines—astrophysics and planetary science—and argues that these disciplines' formative years can only be understood in relation to the cultural marketplace in which they were situated. As Lowell's fight with Pickering suggests, this marketplace was characterized, above all, by new, powerful, transnational forms of mass media. Both these new kinds of astronomy and these new forms of media grew up together from the 1860s, and neither, this book argues, can be fully appreciated separate from the other. I trace this symbiotic history through four detailed case studies, moving chronologically from the advent of astrophysics (1860s and 1870s), through early interest in evidence for a living Mars (1880s), into the heyday of debates over the planet's enigmatic "canals" (1890s to 1900s), and conclude with the slow decline of theories of life on the red planet (1910s and 1920s). As this chronology suggests, Lowell's fight with Pickering comes near the end of my story. Neither man could agree about the physical constitution and appearance of Mars, nor about the appropriate ways in which planetary news should be shaped and circulated through the media. This book explains why these two issues came to become inseparable from one another and why it mattered so much for these astronomers.

The narrative of this book is tightly periodized, being contingent upon two contemporaneous disciplinary moments. The first is the advent of astrophysics and, with it, what much later came to be called "planetary science." It is important to clarify up front that in the period studied here, from around 1860 to around 1910, neither of these appellations were in common use. "Planetary science" was used not at all, and "astrophysics" appeared only occasionally and inconsistently, alongside cognate terms like "solar physics" and "the physics of astronomy." As such, I have adopted in my title an actors' category that encompasses both of these disciplines-in-the-making and which more accurately captures the novel astronomical practices then emerging. Writing in 1885, the commentator Agnes Mary Clerke noted a recent phenomenon, "a so-called 'new astronomy'" that had recently "grown up by the side of the old."[4] Whereas the old astronomy took the movement and mathematical relations of the celestial bodies as its purview, this new astronomy deployed a suite of novel tools and instruments to interrogate the physical constitution of those same objects. Following the canonical spectroscopic work of Robert Bunsen and Gustav Kirchhoff in the early 1860s, astronomers had at their disposal what this pair called an "entirely new method of qualitative chemical analysis" capable of determining many of the elements and compounds present in the sun, stars, nebulae, and planets. As one London observer put it, astronomers could now "come to the chemist" if they "want to know something of the constituents of the heavenly bodies."[5] Concomitant efforts to increase the size and power of telescopes also gave those astronomers hopes of scrutinizing the physical appearance—and therefore physical change over time—of those bodies. Used together, these new techniques opened up entirely new domains of stellar and solar system astronomy, concerned not with position and motion but rather with the physical composition and life history of stars and planets. "The unexpected development of this new physical-celestial science" was, according to Clerke, "the leading fact in recent astronomical history."[6]

From their outset, these revolutionary advances had a powerful public dimension. "One effect of [the New Astronomy's] advent," Clerke observed, "has been to render the science of the heavenly bodies more popular." On the one hand, this was because its results were "more easily intelligible—less remote from ordinary experience" than abstruse positional calculus. Just as important, however, was the fact that "its progress now primarily depends upon the interest in, and consequent efforts towards its advancement of the general public." The new astronomy, she asserted, "depends for its prosperity upon the favor of the multitude who its striking results are well fitted to attract."[7] Clerke, with characteristic insight, could already see what this book aims to recover: that progress in this new astronomy was inextricably bound up with a wider facet of late nineteenth-century culture—the rise of a massive and massively influential public marketplace for science. From its outset, the physical study of stars, sun, and planets was common knowledge, circulating as part of a wide range of scientific content consumed by an increasingly literate and leisured populace. The roots of this marketplace in the English-speaking world have been traced to an "industrial revolution in communication" in the first half of the century. The new technologies of steam printing, rail travel, and telegraphy, in particular, transformed and greatly expanded print culture and, with it, social access to knowledge. These were profound and often controversial transformations, calling into question science's relations to social, political, and religious norms, and stoking contests and controversies over the role of scientific entrepreneurship in the spread of "useful knowledge."[8]

By and large, however, accounts that focus their attention on developments in "popular" science do so through a limited range of genres and consumer products, above all books and periodicals. They often miss, therefore, the central quarry of this book's media focus: a second revolution in print and display, inaugurated in the 1830s in the United States and the 1860s in Britain, that witnessed the rise of a truly *mass* media, principally in newspapers but also in encyclopedias and great exhibitions. This was media consumed not by thousands or tens of thousands but by millions. It was also principally a phenomenon of the transatlantic world, characterized by its speed, cheapness, populism, and, crucially, its novel Anglo-American ambit and sensibility. Though its origins are traced to the "yellow journalism" of 1830s New York, it was only after political and educational reforms in the 1850s and 1860s that its influence was felt in Britain. So at the same moment that commentators discussed the extraordinary revelations of the new astronomy, they also contemplated the rise of a "new journalism." These were rapid, profound, and potentially radical advances in the state of public culture. The affordable and populist print multiplying on both sides of the Atlantic was, one commentator lamented, "bright, racy, trivial, contemptible stuff, which should interest no one of intellectual capacity, and which does interest ninety-nine people out of a hundred." Yet for those more in accord with its progressive, democratizing spirit, this new mass media represented a watershed in public access to knowledge. It was, one advocate wrote, nothing less than "government by journalism," serving as "at once the eye and the ear and the tongue of the people . . . the visible speech if not the voice of democracy. It is the phonograph of the world."[9] Born at almost the same moment, these two revolutionary new practices, the new astronomy and the new journalism, would soon find their paths entwined.

Mars is a particularly good object through which to study this entanglement. My intention is not, however, to present a complete reassessment or systematic reconfiguration of historical accounts of Mars and the various debates over evidence for life there. (These events and their historiography are summarized in the next section.) Rather, my interest is in how more general transformations in astronomical practice that underpinned new accounts of Mars were co-constructed with the transatlantic news economy that discussed and circulated that knowledge. Significant for such an account is the fact that Mars was a challenging object to study and a fascinating object to contemplate and debate. Accounts of Mars from the 1860s onwards focused on the tentative claim that the planet was a living world, somewhat similar to Earth in composition, environment, and topography. These claims hinged on a complex matrix of experimental and visual evidence, incorporating hand-drawn observations made at the telescope eyepiece, spectrochemical analysis, and, eventually, photographic imagery. All of this evidence was then assessed and debated, by necessity, through an unstable prism of terrestrial analogy and biological speculation. This was an entirely new disciplinary terrain, and with it came new problems of how to establish acceptable conventions of practice and discourse. Disagreements about Mars were always also disagreements about how a rapidly changing discipline should be organized, how its new techniques should be deployed, and how and where its diverse practitioners should, or should not, talk about these issues.

As with so many of the great controversies of nineteenth-century science, we must be careful from the outset not to presuppose where authority lay in these debates, nor to preemptively categorize certain aspects of them as orthodox or heterodox based on their eventual fortunes. Assessed on the terms of the actors who populate this book, it soon becomes clear that the possibility of some form of life existing on Mars constituted one of the more pressing questions facing astronomers through the turn of the twentieth century.[10] Important for what follows is the proposition that this question was not tackled or solved within the boundaries of elite astronomical science alone. On the one hand, my study implicates journalists, editors, and public consumers of astronomical knowledge within these debates. On the other hand, I argue that some of those whom we now take to be representative of a stable type of practitioner—the astronomer—actually forged complex hybrid identities that fail to map neatly onto a quintessentially modern distinction between those who "produced" esoteric knowledge and those who "disseminated" it in the exoteric realm.[11] Astronomers were journalists and editors too, eliding practice with communication in consequential ways. The general strictures of astronomy's cultural marketplace—the resources and constraints this public sphere provided—were embedded within and therefore constitutive of the practices of the new astronomy.

The past half century has seen a proliferation of scholarship on the history of Martian study and understanding. As my own approach is not synoptic and as I deploy four detailed case studies within a key phase of this history, it is worthwhile here to briefly sketch the general outlines of this larger story. Setting aside Kepler's important work on the motion of Mars, attempts to study the physical appearance and character of the planet had to wait until the turn of the nineteenth century and the development of large telescopes. Even then, and using the most powerful instruments available, little on the planet could be discerned with certainty beyond an axial inclination, polar caps, and apparent seasonal changes that were all strikingly Earthlike. Yet the continued difficulty of consistently making out fixed features and the suspicion of some observers that Mars possessed a changeable atmosphere that obscured its "areography" below delayed the first attempts to set down a fixed map of the planet until midcentury. These maps, alongside a growing profusion of drawings and written descriptions, soon converged on a general consensus in which the Martian surface was depicted as being composed of darker, greenish areas called "seas" and lighter, ruddy areas called "continents."[12]

That these features might be Earthlike in ways beyond mere topographical analogy loomed large in wider discussions of the planet. As Michael Crowe has chronicled, the first half of the nineteenth century represents the high point for widespread enthusiasm for *pluralism*: belief in a plurality of inhabited worlds in the universe.[13] If natural theological arguments of purpose and plenty had formed the bedrock of such a prevalent belief before 1860, then the experimental results of the new astronomy only served to reinvigorate it after. Antipluralist arguments that posited Earth as a fundamentally unique place were dealt a decisive blow by the evidence of the spectroscope, which found chemical elements and compounds central to the maintenance of life on Earth also present in the sun, stars, and on its nearest neighbors. William Huggins's spectroscopic analysis of Mars's atmosphere, for example, led him to declare the detection of water vapor there in 1867.[14] No astronomer made more hay with this confluence of visual and spectroscopic evidence than the central actor in this book's first chapter, Richard Anthony Proctor. In the same year as Huggins's water announcement, Proctor produced a completely new map of Mars, and then, three years later, he made the planet a centerpiece of his rousing pluralist hit, *Other Worlds Than Ours*. Mars was, in his words, "the Miniature of our Earth," and it therefore presented firm visual and physical grounds for reasoned terrestrial analogy. "Until it has been demonstrated that no form of life can exist upon a planet," he asserted, "the presumption must be that the planet is inhabited."[15]

Within a decade, Proctor's stirring pluralist assessment would receive further corroborating evidence. Working at Milan's Brera Palace observatory in 1877 during an unusually close approach of Mars to Earth, the Italian astronomer Giovanni Schiaparelli systematically studied and remapped the Martian surface.[16] Among his findings was the conclusion that the four large continents on Proctor's map were, in fact, a multitude of islands, separated by a series of narrow strips of water. Working, as was typical, by recourse to terrestrial analogy, Schiaparelli dubbed these new features "canali," in so doing inaugurating the debate that would come to dominate Mars studies for the ensuing half century over the nature and reality of these enigmatic "canals."[17] At play in this great "Martian canal controversy" were a wide range of interlocking questions: Did these "canals" actually exist as delineated? Did they periodically appear, disappear, or even double into parallel sets? Were they channels of water? Or areological fissures? Or bands of vegetation? Or optical illusions? Or the manufactured edifices of an intelligent race of Martians? Many of the details of this debate will feature in the chapters that follow. At this stage, it is enough to say that the nature of the canals themselves followed a clear representational arc, in which their geometric certainty increased over time, as Schiaparelli's relatively few large canals multiplied through a succession of discoveries and resightings by a range of observers (see figure I.1). By the turn of the century, the canals had emerged as a comprehensive network of narrow, straight features, inscribed into a small set of precise, abstract maps that, as K. Maria D. Lane has persuasively shown, "eventually became powerful cartographic icons that were viewed as indications of intelligent Martian life."[18] On this telling, the network seen on Mars was simply too geometric and rational to be anything other than artificial, a point extrapolated by Lowell in particular into a comprehensive theory of a planet-wide irrigation system, manufactured by industrious Martians in the face of an arid and dying planet. It was these ideas that so excited astronomical and public interest in the planet and that so agitated more skeptical observers like Edward Pickering.

The notion that Mars was covered in a canal network of some kind did not entirely die out until the mid-1960s, when NASA's Mariner 4 mission finally secured close-up photographs of the planet's surface. During the ensuing five-plus decades, historians have developed a rich and varied account of the planet as it was understood before this time, as a body of enigmatic features and questionable habitation. Within this historiography, three distinct eras can be readily determined. The first emerged in the immediate aftermath of the canal's final demise and represents planetary astronomy's own discipline-building efforts to explain—or rather explain away—an episode that these practitioners suddenly found deeply embarrassing, if not an active threat to their reputations. In these accounts, therefore, emerges the first sense that the canals and broader ideas of Mars as a living planet represented nothing more than an unimportant mistake, liminal to the workings of astronomy proper. Faced with a recognition that the trouble over Mars could not be conveniently ignored, this literature instead

Figure I.1. Globes of Mars, 1873–1913. Considered in chronological sequence (*left to right*), these globes demonstrate the changing representation of Mars by astronomers over the period of this book's study, moving from an Earthlike planet of continents and seas to a desert planet dominated by a comprehensive network of geometric, interconnected canals. *From left:* Nine-inch globe of Mars by Malby of London, 1873. Commissioned by Captain Hans Busk the Younger and based on the 1867 map by Richard Proctor (see figure 1.1), this was the first mass-produced globe of the planet. Eleven-centimeter globe of Mars, published by Camille Flammarion and made by E. Bertaux, Paris, 1884. This is the first globe to show Schiaparelli's "canali," which here retain their naturalistic shape as broad channels between broken up sections of continent. Ten-centimeter globe of Mars, published by Louis Niesten and made by J. Lebèque, Brussels, 1892. As the number of reported canals increased, they also narrowed and their edges began to straighten. Fifteen-centimeter globe of Mars, published by Camille Flammarion and Eugène Antoniadi and made by E. Bertaux, Paris, 1898. Flammarion became a keen advocate for evidence of life on Mars, and his own maps of the planet shifted the representation of the canals toward their being a comprehensive network of narrow features on an otherwise dry landscape. Eight-inch globe of Mars by Ingeborg Brun after the maps of Percival Lowell, Svendborg, Denmark, 1913. Lowell's maps of Mars represent the apotheosis of the artificial canal theory, and would come to be the dominant representation of the planet at the start of the twentieth century. Image © Whipple Museum of the History of Science, Cambridge: Wh.1268; Wh.6622; Wh.6625; Wh.6238; Wh.6211.

recast it, anachronistically, as a salutary tale of astronomy gone wrong, in which the advocates for an inhabited planet were assessed and characterized on the terms of the Space Age as nothing more than outsider "popularizers." Such accounts are actively harmful to our understanding of the subject, and I return to their cause and effect in the conclusion.[19]

From the late 1970s, a number of historians of astronomy began to take Mars and its canals seriously, and in their work we find the first critical accounts of debates about the planet's physical constitution. These histories make clear that in the wake of the canals' discovery and amid continued widespread fascination with the question of plurality, a diverse range of actors, including the public at large, began to earnestly and vigorously debate the evidence for and against Mars's possessing some form of organic life—possibly highly evolved and intelligent life. Front and center in such accounts is the man typically understood as the driving force behind the canal controversy, Percival Lowell. Portrayed as idiosyncratic, energetic, and hugely successful at capturing a large public audience, Lowell's advocacy of evidence for a living Mars—centered upon his own observatory's maps and an accompanying series of very successful books—dominates this historiography.[20] As helpful as this work is, therefore, it propagates a problem I have attempted to avoid in this book, that of "Lowell mania." Lowell was, in truth, a relative latecomer to debates over Martian life, establishing his observatory in Flagstaff, Arizona, over a decade and a half after discussion of the canals had begun. Furthermore, such focused attention on Lowell perhaps distorts his role and those of his key rivals and risks dividing the debate into a "popular" Lowell camp and a "professional" anti-Lowell camp. The result has been a sense that the work of Lowell's critics represents something like a singular, coherent, and rational response and thus is not itself worthy of close analysis.[21] Of the four case studies that comprise this book, only the last features Lowell, and it does so mainly from the perspective of his opponents, who, I argue, were just as invested in attracting popular audiences as their Flagstaff colleague.

My own account most directly draws on and augments work from the third era of Mars studies. The last twenty years have seen the emergence of a new cultural history of the red planet that has integrated and synthesized scientific, social, and political narratives, greatly invigorating our wider understanding of Martian astronomy. Literary, ecological, geographical, and visual aspects of the debate have been fruitfully explored, explicating seemingly eccentric claims about Mars through a sociocultural analysis of scientific practice, representation, and reception. This literature has shifted accounts of Mars away from the more timeless, philosophical framework of the "plurality of worlds" question, recontextualizing the planet as a significant facet of late nineteenth-century culture. A common theme across these works is the ability of Mars to act as a cultural mirror apt for the reflection of scientific, social, and political narratives and concerns. Mars was an object of study, but it was always also a site of projection for terrestrial affairs, which ranged from imperialist conceptions of the "other" to fears of ecological decay and entropic heat death.[22] This historiography has greatly advanced our understanding of both Mars and the era's wider culture, and this book draws on and reinforces many of its findings, while also diverging from it in several important ways. I do not, for example, share these studies' often keen interest in the science fiction of the era. This topic has already been extensively treated, and—with some notable exceptions—it falls outside this book's focus on the implication of media within astronomical practice itself.[23]

Placing media at the center of my account does not make this a book about popular science or about popularization. This is not to deny the immense significance of the public sphere in what follows. It is, rather, to foreground a key element of my general argument: that media matters in an assessment of news from Mars precisely because of the impossibility of separating out the popular from the professional in this story. Central to this book's account is the now well-recognized point that categories such as "professional" were very much still in the making at the turn of the twentieth century. To use them unthinkingly, therefore, is to risk freighting into our historical understanding a suite of essentially modern assumptions about status, roles, and expertise and in so doing to very obviously beg the question.[24] This is a particularly important point for the history of early astrophysics, given the still common assumption that astronomy by 1900 had witnessed the field's bifurcation into "amateur" and "professional" camps.[25] To take such a divide seriously is to risk imposing onto an unusually unsettled period of disciplinary reorganization the categories established after the fact by the victors of this struggle.

Misconceptions about how actors used media are one of the central features of this kind of anachronism. Especially when it comes to talking about life on Mars, it is all too easy to slip into a modern assessment of certain skeptical actors as inherently professional and pit them against an opposing force of mere popularizers, with the latter group making up for their lack of astronomical expertise with a countervailing wealth of skill as mass media manipulators. A now large body of recent scholarship on Victorian science in the public sphere has cogently called into question just such a set of categories and divisions. In so doing, these accounts have effectively decentered historians' view of where scientific authority lay during the nineteenth century.[26] The case studies explored here build on and corroborate this work, and in what follows I place particular stress on the ways in which supposed dichotomies between popular and professional are actually undermined. Widely read champions of life on Mars were also major players in the senior hierarchy of the Royal Astronomical Society. August and skeptical mathematical astronomers were also authors of cheap books and encyclopedia entries on Mars. These seemingly topsy-turvy relations of expertise and public science did not necessarily endure, but their eventual demise makes them no less central to the story at hand.

My account is as such an exploration of a fluid and conflicted moment in disciplinary history, at a time when norms of practice were in flux and categories such as "amateur" and "professional" were rarely used. The challenge becomes explaining how this instability was caused, what its effects were, and how, eventually, it came or did not come to be resolved. The answers lie, I argue, in the entanglement of new forms of public discourse with new forms of astronomical practice that were themselves unusually public. By the end of the period studied here, some of this uncertainty had been removed, some of it by processes that we might fairly call professionalization. But we need to understand the contingent and situated processes through which this occurred before we can begin to safely apply such categorizations onto histories of astrophysics and planetary science.[27]

In seeking to recover actors' understandings of expertise and its relations to media, my analysis deliberately shifts attention away from any broadly construed public sphere, focusing instead on a more limited set of individual practitioners active within it. My intention is to pursue an analysis of media in science rather than science in media. This means turning away somewhat from the streets and raucous showrooms of "spectacular science" and returning to the observatory, study, and lecture hall to recover the roles played by media in science at an essentially disciplinary level.[28] This approach enables me to scrutinize the particular, situated sociology of roles at play in disputes over evidence for life on Mars. Disagreements and assertions regarding astronomical technique and disagreements and assertions regarding appropriate forms of astronomical discourse were, I suggest, both related elements of the same set of social actions. In both cases, practitioners were working to establish priority, defend findings, and redraw disciplinary boundaries. To accomplish these goals meant drawing on material, literary, and social resources from the wider cultural spheres within which these actors positioned themselves, and this maneuvering was shaped, therefore, by a practitioner's status, income, and favored techniques, as well as their various publics' expectations. To trace the progress of these moves is to study not the inevitable development of advancing professionalization or disciplinary regimentation, but rather a range of subtle and multifaceted contests over authority and hierarchy within the scientific community.[29]

In the case of Mars, astronomy's sudden plurality presented a range of apparently viable options for how planetary studies should be publicly presented and professionally certified. In the first chapter of this book I recover one such option, which I dub "imaginative astronomy," an egalitarian and anti-elitist model that predicated successful practice on its financial success in mass media marketplaces. Imaginative astronomy was, I argue, a guide for disciplinary practice that was hugely influential in its time but that has since been lost to the historical record through its ultimate defeat at the hands of less social alternatives. Its recovery suggests that the eventual concretization of only certain norms of practice should not obscure from us the various ways different practitioners worked to forge a range of identities for themselves and, with them, alternative principles for their discipline. Mars became a crucial site for the elaboration of such identities and models.

A key concept within this analytical framework is that of mass media. Astronomers' efforts to forge viable identities required the marshaling of allies and the projection of knowledge claims, work for which genre proved to be pivotal. The ways that new knowledge appeared—chronologically, geographically, and stylistically—affected the reception of that knowledge and therefore its meaning and impact. How practitioners responded to the expectations and actions of diverse consumers of astronomical content must, therefore, be understood as a trajectory of action shaped by these marketplaces.[30] Central to my account is the claim that it was *mass* media rather than esoteric specialist journals or private correspondence networks that proved particularly important for the progress of news from Mars. As such, this category is used here specifically to denote genres of mediation that had a very wide impact, reaching large audiences and penetrating deep into the public sphere. This is not to suggest that mass media can be taken as a value-free or neutral category—"the masses," after all, was one way in which turn-of-the-century intelligentsia dismissively characterized a newly educated reading public that they neither understood nor cared for. But as an overarching category for my account, it nonetheless serves a useful purpose, foregrounding as it does the era's cultural democratization and the diversity and sheer quantity of new forms of communication and audiences.[31] Its use also leaves me free to take note of the ways in which my own actors deployed the categories of "popular" and "popularization."[32] Most importantly, the phrase "mass media" captures the fact that the debates under analysis progressed during a period in which media established itself on a truly massive scale internationally, making it pervasive within the astronomical sciences. Telegraphically networked newspapers that reached audiences in the millions, encyclopedias that sold in the hundreds of thousands, expositions that were experienced by tens of millions, lecture tours that visited hundreds of towns, and cheap periodical literature with circulations in the tens of thousands all mediated novel knowledge claims about planetary astronomy, and all were therefore part of the working world of astronomers.

My contention is not that Martian astronomy was merely "mediatized" in this era, a process that suggests the external agency of a nonscientific reportage appended to the astronomy itself.[33] There was no necessary disjuncture, I argue, between the scientific event and the media that covered it. Often the journalist was also the astronomer. Time and again, my story indicates that mass media were—or at least could be—embedded within astronomical practice itself. As a result, the (always contested) place of mass media *in* astronomy was integral to the shape and fortunes of the various fights over validity, authority, and hierarchy occasioned by the advent of the new astronomy. The actors within these debates themselves recognized this fact, enabling me to trace the specifics of these astronomers', journalists', and editors' defenses of and attacks upon various forms of mediation. These arguments, as discipline-defining acts intended to delineate right practices and unsuitable "others," were themselves necessarily mediated processes. This was not a question of scientific versus popular media, but rather a much more subtle and intricate question of genre and its relations to practice. My account therefore highlights the fact that critics of certain types of popular astronomical discourse were often not themselves any further removed from mass media but were, rather, closer to different types of mass media—hence my account's concern with not only mass-circulation periodicals and books but also newspapers, globes, public expositions, magic lantern lectures, and encyclopedias.

One salient consequence of a media-focused history of astrophysics and planetary astronomy is the attention it draws to both the temporality and geography of the movement of knowledge. Time is significant for my account because Martian observations and the rhythms of mass media were consonant with one another through their mutual seriality. Serial modes of organization, production, and communication loomed large over nineteenth-century science, and the periodic nature of such practices was a central feature of news from Mars. Mars approached Earth at distinct time intervals (see figure I.2), meaning that new observations and claims about the planet were disseminated and consumed in discrete, concatenated chunks. Fascination with the red planet waxed and waned according to its periodic orbital cycles, and people's responses to the latest astronomical research changed and progressed over these serialized time intervals. Mars's ability to generate news was consonant, then, with the rhythms of the era's serialized platforms of communication, such as the newspaper, the periodical, and the lecture, all of which "defined knowledge as a material commodity distributed, consumed and disposed of on a regular basis."[34] This seriality, I suggest in chapter 3, propagated a form of "event astronomy" in which the meanings of novel knowledge claims emerged from the context of their specific spatial and chronological movement.

A signal feature of this event astronomy was its transatlantic scope. The new technologies that underpinned the emergent mass media of the late nineteenth century—above all telegraphy—transformed both the extension and the social possibilities of communications practices.[35] Disciplines like astronomy whose new, contested epistemologies were both public and organized in part through mass media therefore underwent changes consequent to these new social encounters and cultural developments. In the English-speaking world, one key element of this transformation was the post-1850 emergence of a common transatlantic form of journalism defined by its speed, boldness, populist spirit, and mass appeal. Crucial in what follows is the observation drawn from the history of journalism that such changes were essentially transnational, but only in the restrictive sense of being Anglo-American. Underpinning the convergence of the practices and norms of English-language journalism, it has been argued, was "a common framework of democratization and of joint cultural formation."[36] As this book traces the implication of this Anglo-American media within

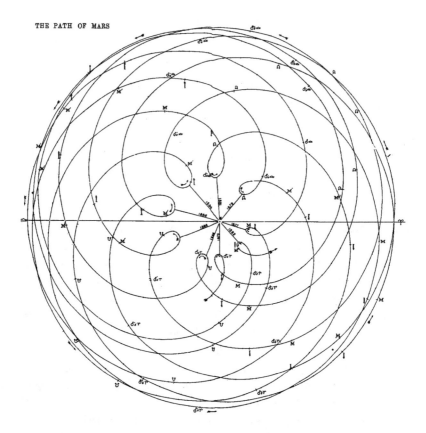

Figure I.2. This chart shows the path of Mars relative to a stationary Earth (at center), from 1877 to 1892. Mars could only be scrutinized at brief periods of close approach—called "oppositions"—that occured roughly every two years (as indicated by the nine loops near Earth in this chart). The quality of these oppositions also varied, with the best, called "perihelic oppositions," occurring only every fifteen or sixteen years, here shown by the closest approaches in 1877 and 1892. Martian study and debate therefore tended to follow the episodic rhythms of the planet's path. Richard Proctor, "The Path of Mars," *Knowledge* 1 (March 24, 1882): 452.

astronomical practice, it follows actors almost exclusively from Great Britain and the United States (though their actual working sites were, as we shall see, somewhat more geographically diverse). Indeed, one striking feature of this book's four case studies is the dominance of the English language in all of them. French astronomer Camille Flammarion, for example, who was spectacularly successful in his own country as an author on Mars and extraterrestrial life, only began to make a significant impact on transatlantic news from Mars when his works were (somewhat belatedly) translated into English, most prominently by the Paris edition of the *New York Herald*. This primacy of English in transatlantic coverage of Mars reiterates the overwhelming importance of mass audiences in my story. Languages prevalent among working astronomers, such as French and German, made little headway in a debate dominated by a globally networked English-language media.[37]

Neither the telegraph nor mass media altered astronomy's social relations and practices overnight, however, and their impact needs to be traced over time and—as befits our subject's serial progress— in chronological order. We begin therefore in the period between the early 1860s and the early 1880s, at the dawn of the age of Martian intrigue. Chapter 1 chronicles the rise of the new astronomy, assessing the key role played by Mars in this nascent subdiscipline's disputes and contests and arguing for the central importance of a new transatlantic media marketplace in early debates over life on the planet. New Anglo-American journalistic ideals of egalitarianism and anti-elitism, it is suggested, underpinned a powerful new model for astronomical practice—imaginative astronomy—that, though relatively short lived, had a long-term impact on conceptions of the red planet. Chapter 2 then considers a significant geographical shift in astronomy during the 1880s and 1890s, consequent to reorientations of the place of the observatory in an age of internationally networked telegraphic news distribution. This move positioned certain privileged locations in the American West as new, exceptional sites for Martian study and implicated these sites within a marketplace of newspaper astronomy. Chapter 3 explores one episode in this new marketplace, the "great Mars boom" of 1892, and illustrates how the implication of newspaper coverage within the working practices of certain observatories shaped the emergence of a canal-focused narrative for Mars. Chapter 4 then presents one significant response to the extraordinary speed and reach that came to typify both the power and the potential problem of this newspaper astronomy. In the early 1900s, I suggest, it was the combined scope, authority, and permanence of encyclopedias that made them an important medium through which rival disciplinary claims could be presented. My conclusion, finally, considers the nature of the slow decline in interest in a living Mars during the first half of the twentieth century and addresses some of the methodological lessons that can be drawn from my overall account. But first, we must start at the very beginning, with the shock of the new.

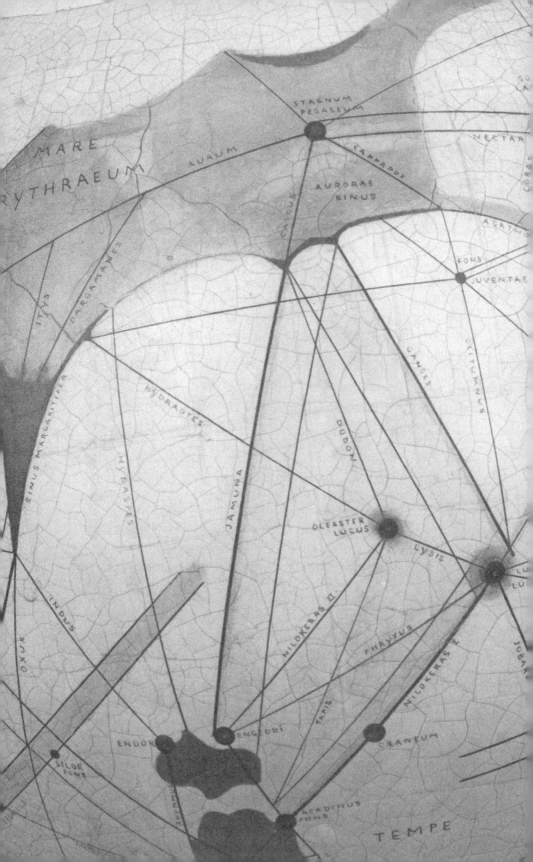

WRITING ON MARS

1

Imaginative Astronomy
and the New Journalism

It started with a letter. The *Times* of London, April 13, 1882, page 12, a letter to the editor. For many in the English-speaking world, this letter—or subsequent reprints and reports of it—was their first encounter with the idea of canals on the planet Mars. The letter had been written by the astronomer Richard Anthony Proctor, something of a celebrity in the world of science and a name likely to have been familiar to many of the *Times*'s readers. In an era when astronomy commanded huge popular interest, Proctor was by some distance the subject's most widely read authority. His "name as an expositor of science," one contemporary noted, "has become a household word wherever the English Language is spoken."[1] Proctor's letter responded to news relayed by his colleague Thomas William Webb that "minute, straight black or dusky bands" first observed on Mars from Milan by "the eminent astronomer Signor Schiaparelli" during the perihelic opposition of 1877 had recently been resighted. Remarkably, Schiaparelli's new observations of these "canals" included "the duplication of these dark streaks by the addition of parallel lines of similar character and length in no fewer than 20 instances, covering the equatorial region with a strange and mysterious network."[2] Proctor replied to Webb with a tentative public endorsement of these features, noting their presence in the sketches of Mars in his own possession, made over a decade earlier by his late friend William Rutter Dawes. But he also urged caution that "we . . . not too hastily assume that these are real features of Mars." Should such a network be proved to exist, however, Proctor concluded that "we should by no means be forced to believe that Mars is a planet unlike our earth; but we might perhaps infer that engineering works on a much greater scale than any which exist on our globe have been carried on upon the surface of Mars. The smaller force of Martian gravity would suggest that such works could be much more easily conducted on Mars than on the earth, as I have elsewhere shown. It would be rash, however, at present to speculate in this way."[3]

To a modern reader, this letter may seem hard to fathom. Considered without an understanding of its particular context of production, it comes across as sensational and paradoxical, riven by incongruent styles and arguments. Even when read more sympathetically as a product of Victorian popular science, there are a range of pluralities in its content that demand closer scrutiny and explanation. First, the letter presupposes at least the possibility of a plurality of inhabited worlds. By 1882 this idea had, in fact, been a central feature of astronomical and theological discourse for several centuries, with Mars serving as a key site through which the vexing question of Earth's uniqueness—or not—could be probed and puzzled over. In this regard, Proctor's letter can be read as an important intervention into an already long-standing debate—one that would, by the turn of the century, come to be dominated by its principal subject, the Martian canals.[4] Yet we can hardly take Proctor's letter as an origin point for the advocacy of Schiaparelli's canals as the manufactured edifices of intelligent Martians, given its cautious and equivocal endorsement of the evidence presented.

Indeed, assessed with modern eyes, another completely different kind of plurality becomes evident: that of Proctor's multidisciplinary approaches. On the one hand, Proctor very much worked in a register that we might expect from a modern "professional" astronomer, offering a natural skepticism that urged against jumping to conclusions based on scant astronomical evidence. Yet, on the other hand, he rapidly switched from this careful tone to a bold style of playful speculation and imaginative conjecture—a style that we might more readily expect from a "popularizer" than from an eminent fellow of the Royal Astronomical Society. Underpinning this plurality of practices, therefore, is another, subtler plurality: of genre and, with it, of identity. Proctor corroborated Webb's tentative report through a letter to the *Times*, citing private observations he judged preeminent and speaking of calculations that he had "elsewhere shown." His authority in these claims derived from a reputation won both by his complete re-mapping of Mars and through his many successful popular books and lecture tours that described and analyzed Earth's outer neighbor. Who Proctor was—astronomer? popularizer? man of science? journalist?—can be very tricky for us to pin down, especially if we anachronistically assume that he had to be only one of these things. Yet if we instead accept that he might in fact have been *all* of these things, then we are left with the challenge of unpicking where Proctor's letter sits within astronomy's disciplinary landscape, how his audiences would have read and understood it, and what, in turn, this tells us about Mars, canals, and Victorian debates over intelligent extraterrestrial life.

This chapter is an attempt at such an unpicking. In taking Proctor as its central actor, it presents a thorough reassessment of an astrono-mer who was, as Michael Crowe has noted, for the last three decades of the nineteenth century "the most widely read writer on astronomical subjects" in the English-speaking world.[5] But in placing Proctor's di-verse work at the heart of a new disciplinary landscape for astronomy, this chapter argues against reading Proctor as merely an author. Cate-gories such as "popularizer" and "writer," I argue, greatly underdeter-mine who Proctor was and how he worked. This is no less true for his colleagues, and in what follows I read a wide range of works in astron-omy as forms of media, each with their own particular genres, mar-ketplaces, and audience responses. This is to recognize that astronomy in the period covered here—from the early 1860s to the mid-1880s—was a discipline embedded within a particular transatlantic cultural moment.

The chapter is composed of two roughly equal parts. In the first four sections, I recover a form of astronomical practice that I dub "imaginative astronomy." Proctor developed this practice, I suggest, as a challenge to the working methods of two better-known factions in the British astronomical community—the old guard of "gentleman" positional astronomers, and the younger upstart "scientific naturalists." In characterizing the fights that ensued over Proctor's new model and the place of Mars and mass media in these fights, I attempt to define what made imaginative astronomy unique, special, and powerful. In the final two sections of the chapter, I shift my analytical framework away from astronomy to think instead about journalism. Here I argue that a context of transatlantic, populist, democratic mass media explains the particular form, as well as the spectacular success, of Proctor's work. Imaginative astronomy is shown to emerge from both cutting-edge astrophysical research and the resources and constraints provided by an innovative genre of Anglo-American media—the "new journalism." Reassessing this period of astronomical labor on these terms enables us to arrive at not only a more complete historical understanding of Proctor's seminal letter to the *Times* but also of its wider context, of astronomical practice and its reception in an era of news from Mars.

NEW ASTRONOMIES

The path to understanding Proctor's letter begins at the start of his career as an astronomer. Significantly, he entered the discipline in the early 1860s, at a time of turbulent upheaval for the field. Robert Bunsen and Gustav Kirchhoff's pioneering work in the spectroscopic analysis of celestial objects had inaugurated an entirely novel research agenda, with London, Proctor's home, soon becoming a major center for this "new astronomy." Central to this reconfiguration of disciplinary technique was a reorganization of work spaces. As astrospectroscopist William Huggins later noted, it was at this moment that observatories "began, for the first time, to take on the appearance of a laboratory." Batteries, induction coils, Bunsen burners, vacuum tubes, and an assortment of chemicals joined cameras and spectroscopes in the service of multifarious experimental trials. With them the orderly rhythm of observation by eye and ear was disrupted by complex assemblages of laboratory labor, itself contingent on often fragile links with chemists, photographers, and instrument makers.[6] The early results of this work were striking in their implications for the scope of astronomy's disciplinary purview and philosophical extension. Huggins and William Miller's seminal 1864 paper "On the Spectra of Some of the Fixed Stars" demonstrated the fundamental commonality between chemical elements found on celestial objects and those found on Earth, including "some of those most closely connected with the constitution of the living organisms of our globe." This, the pair concluded, contributed no less than "an experimental basis on which a conclusion, hitherto but a pure speculation, may rest, viz. that at least the brighter stars are, like our sun, upholding and energizing centers of systems of worlds adapted to be the abode of living beings."[7] Astrophysics therefore promised profound new insights into one of the oldest debates in celestial studies, over whether or not there existed a plurality of living worlds in the universe. Pertinent too for these discussions were questions of the structure, constitution, extent, and history of the universe. Here the spectroscope provided new evidence on the constitution of nebulae and the line-of-sight velocity of stars, while another novel chemical technique—photography——enabled enhanced photometry, astrometry, and sunspot surveillance. Taken together, then, these novel techniques of the new astronomy enabled a greatly expanded disciplinary domain. In so doing, however, they also massively destabilized astronomy's power structures and management regimes. Who should have control over these new interdisciplinary practices, who should police their norms and boundaries, and how and where novel evidence and claims should be made were all more or less open questions.[8]

Proctor's entry into this unstable world was idiosyncratic, to say the least. The son of a solicitor, after studying mathematics at Cambridge he had entered business and studied for the bar before a bad investment in a New Zealand bank in 1866 left him and his family in heavy debt. Forced to seek a living quickly, he turned to writing about astronomy, a practice that he had begun as a hobby and now intended to pursue as a full-time career. Staggeringly prolific, Proctor soon carved out a role for himself as both an active contributor to the elite London astronomical scene and as a financially self-sufficient author servicing a booming demand for astronomical news and affordable educational and scientific writing.[9] Although he had no observatory or institutional base from which to work, Proctor's early work was considered impressive enough to get him elected a fellow of the Royal Astronomical Society (RAS) in 1866, and within a few years he was serving as an honorary secretary with responsibility for editing its *Monthly Notices*. Proctor's place in the disciplinary landscape of new astronomy has, however, been mostly overlooked, not least because of his extraordinary success as an author of very popular works. Characterizing Proctor as merely a popularizer, however, risks missing the considerable reputation and authority his diverse labors brought him. As Bernard Lightman has noted, "to the Victorian reader, Proctor was as much an exemplary scientist as Faraday, Humboldt or Darwin."[10] Like those men, such authority was won through a skillful combination of demonstrated expertise and engaging popular exposition. Proctor studied a wealth of published data to construct new charts and theories on star distribution and the large-scale structure of the universe, and reviews, for example of his Royal Institution lecture on star drift and nebulae, typically noted his skills in both professional and popular realms: "Mr. Proctor has triumphed, not only as an original observer of persevering enterprise and surprising acuteness, but as one of the best popular teachers of astronomy which the country has produced."[11]

Such a position and reputation gave Proctor a powerful but contentious footing in the venerable RAS. Indeed, his rapid election can perhaps be ascribed to the society's sudden instability and confusion in the heady days of the (very) new astronomy. Certainly, the RAS Council that Proctor joined was already extremely divided, and he took an active role in its affairs during an era of "factious" and "unseemly" infighting.[12] The novel experimental techniques of the new astronomy were received uneasily by the discipline's traditional power elite, a group centered on the "Greenwich-Cambridge axis." At this group's core were the Astronomer Royal George Biddell Airy, his allies at Cambridge University James Challis and John Couch Adams, and Airy's staff at Greenwich. Together these men formed a close-knit group of "gentlemen of science," almost exclusively trained in Cambridge mathematics. From their secure institutional bases, this small cohort of "business astronomers" had commanded the RAS since its foundation in 1820, and they had worked hard to establish the Newtonian celestial mechanics of precision positional astronomy as their discipline's dominant practice.[13] In 1864 the aged naval officer and ex-president of the RAS Admiral William Henry Smyth summed up this group's position succinctly when he expressed the hope that "Chemistry . . . not be exerted among the Celestials to the disservice or detriment of measuring agency, and this I hope for the absolute maintenance of Geometry, Dynamics, and pure Astronomy."[14] Statements like this were pointedly directed at a new rival faction then emerging in the discipline— a group of practitioners energized by the new tools of astrophysics. Maneuvering around and increasingly against Airy and Smyth's old guard were a younger generation of experimentalists, centered on the precocious astrospectroscopist J. Norman Lockyer and his close ally Warren De la Rue. Little interested in Greenwich's painstaking angular measurements and laborious calculation of navigational ephemerides, this new guard promoted an agenda of professionalized new astronomy, bolstered, as we will see, by a rich array of new publications and audiences.

Proctor, quite remarkably, never fell in with either of these major camps. As a young and precocious neophyte in a discipline that offered limited scope for direct paid employment, it's hardly surprising that he was deeply skeptical of the Greenwich-Cambridge axis, an entrenched elite that controlled the few salaried positions in the field. Always argumentative and never keen to bow to authority, Proctor loathed the hierarchical "military" model of positional astronomy and resented that its senior figures received public remuneration yet remained "perfectly free from all possibility of criticism."[15] And as a cosmological theorist and an author, he had little time for the navigational ephemerides that were the field's practical yet boring outcome. Though Cambridge trained, Proctor also lacked the social position and therefore shared norms of the Greenwich-Cambridge gentleman practitioners. Like some others of his younger generation, he paid little heed to the expected decorum of London's establishment elites and largely ignored the unspoken codes that had silently governed standards of discourse and hierarchies of authority in the RAS since its inception.

Yet, more surprisingly, Proctor also refused to fall in with the Lockyer crowd. These young men constituted a growing and organized group, predominantly from lower- and middle-class backgrounds, who, like Proctor, were both ambitious and keen to make their scientific work pay. Historians have typically categorized these men, who organized through the now-famous X Club dining fraternity, as the "scientific naturalists," a cohesive and pugnacious group of modernizers that lobbied in favor of meritocratic, institutionalized research, state support for science, national scientific education, the abolition of aristocratic patronage, and the establishment of a naturalistic worldview that eliminated the interference of theology in all scientific matters.[16] Though seemingly a much better fit for the young Proctor and though logically a viable source of support and amity, this group too soon found itself the target of criticism and invective from Proctor in debate and in print.[17] On the surface, one can immediately read into these disputes fundamental disagreements over certain specifics of their science. Proctor had a reputation to build and maintain, pitting him against colleagues with much better access to the costly tools of astrophysics. Close scrutiny and contentious theorizing directed against those nearest to him in stature and methodology was, therefore, a high risk, high reward strategy for forging a professional identity.

Typical was Proctor's disagreement with Lockyer over the nature of the solar corona. Lockyer's reputation hung almost entirely on his novel work on the chemistry of the sun, and Proctor used his increasingly successful books and articles to mock his colleague for suggesting that the corona was an effect of light scattering in Earth's atmosphere.[18] But scratch beneath this surface of astrophysical antagonism and it quickly becomes clear that these arguments were always about more than just solar physics. In an era when new marketplaces empowered a diverse array of actors to delineate and promote a range of definitions of what science was and who could participate, Proctor took a position against the *method* of the scientific naturalists. Lockyer and his cohort pursued the wrong sort of modernization, he argued, because they remained loyal to an elitist creed of scientific exceptionalism. Rather than trying to reform or break away from privileged and cloistered institutions, the Lockyer crowd merely sought to establish new forms of elitist control over science. Seemingly iconoclastic but always backed by a vast and easily overlooked cohort—his loyal readers—Proctor pressed the case instead for a combatively egalitarian, anti-elitist, anti-institutional, and unashamedly populist form of astronomy.

PROCTOR AT WAR

A couple of early salvos are telling. With little care for propriety, in 1869 Proctor launched a headlong attack against the RAS's most eminent member, publicly criticizing the Astronomer Royal's preparations for the nation's forthcoming transit of Venus program. George Airy had been tasked with selecting sites for five widely dispersed field stations from which the passage of Venus across the sun in 1874 and 1882 could be accurately timed in an effort to better determine the distance of the sun from Earth. With what would prove to be a typical blend of intellectual grandstanding and a flair for publicity, Proctor accused Airy of using dodgy admiralty maps and erroneous calculations in his plans for situating the stations. Airy, no doubt disinclined to become embroiled in an argument with such a junior member, at first ignored these myriad public attacks and attempted to keep his response in private channels. In the lead-up to the departure of the first expedition, however, the matter exploded onto the national stage when a scathing anonymous article in the London *Spectator*—certainly written by Proctor himself, a fact likely recognized by many at the time—warned that Britain would "suffer serious discredit" unless the "author of the mistake" did not immediately correct his "astronomical blunder."[19] Coverage in the *Times* followed, letters from the secretary of the admiralty were sent to Airy, and, most embarrassingly for the Astronomer Royal, questions on the matter were raised in Parliament. Proctor by this point was already quite comfortably the most widely read authority on astronomy in Britain, and he pressed home his attack in two popular books on the coming transits—including chapters entitled "Corrections of the Astronomer Royal's Statements" and "Risk of Absolute Failure."[20] Airy likely knew that Proctor was also the anonymous author of various letters in the press, and he wrote witheringly to Proctor explaining his silence on the matter: "I do not think that any good comes from letters of this class, and I should not imitate them."[21] But Proctor ultimately forced Airy's hand through a direct appeal to the Board of Visitors to Greenwich, compelling the Astronomer Royal to negotiate with the admiralty for revised site locations for the transit team transported by HMS *Challenger*.

Proctor's methods angered many in his discipline, not least because it was evident that he had abused his power as editor of the *Monthly Notices* when he had published a supplement that contained nothing but his own transit articles and maps. But the greater scandal had been his behavior toward his colleagues, which had so brazenly run roughshod over expected norms of society debate and which had so publicly aired a sensitive scientific matter. However, although sections of the council were agreed on the need to censure Proctor, the matter was greatly complicated by the almost total chaos that this body had descended into over the course of the transit farrago. While Airy had been defending himself against Proctor's pen, a separate storm had arisen within the RAS that had divided the council along its most fragile line, between positional astronomers on one side and their upstart astrophysicist rivals on the other.

Ostensibly, this dispute hinged on the vexing question of state funding for science. Greenwich had long been financed by government stipend, but the scope for state support of astronomy outside the walls of the Royal Observatory was extremely limited. In April 1872 one of Lockyer's close allies, retired army surveyor and inspector of the empire's scientific instruments Alexander Strange, began to push the RAS to endorse a new, separate, state-funded observatory dedicated to "the Physics of Astronomy." To the Greenwich old guard, it seemed rather obvious that Strange's cohort was maneuvering to attack the primacy of the Royal Observatory by establishing a rival national observatory, with Lockyer inevitably at its head. Even Proctor could sympathize with Airy on this matter, writing to warn his adversary that Strange was evidently angling to "set up a certain junior member of our council as a sort of 'Astronomer Royal for the Physics of Astronomy.'"[22] When the topic finally came before the RAS Council, Strange's motion contentiously failed, only stoking the rancor now growing between what had become a three-sided conflict. Matters reached a boiling point at the RAS's November meeting, when De la Rue, Strange, and Lockyer all resigned from the council. Airy urged Lockyer to reconsider for the sake of stability in the society, but Lockyer's reply was blunt: "Permit me then to tell you at once in confidence that my primary reason for quitting the Council is the offensive manner in which Mr. Proctor is conducting himself towards me. Week after week in more or less obscure journals which as Editor of *Nature* I must see I find myself attacked by one who takes good care to advertise himself as 'Honorary Secretary of the Royal Astronomical Society.'"[23]

More drama soon followed. At the same meeting in which Lockyer departed the council, his friend Charles Pritchard proposed him for the society's prestigious Gold Medal, nominated in conjunction with Jules Janssen and Lorenzo Respighi for their work in solar physics. Against this slate one other nominee emerged: Proctor, nominated by his own allies within the society in recognition of his work on star distribution charts and star drift.[24] In this contest, Airy seems to have abstained from supporting Lockyer, but he fell in vehemently against Proctor. The stakes, as Airy saw them, could not have been higher. When John Couch Adams wrote to the Astronomer Royal asking his opinion of Proctor's candidacy, Airy responded with a withering assessment of his merits as an astronomer: "A new Spirit has come into the Council," he declared, "whom at length I refused ever to transact business again. *Make him President, and the Society will go to the D.*"[25] Though no doubt in part born of personal animosity, it is significant that this criticism implicitly links Proctor's populist approach—his "new spirit"—to his unsuitability as a recipient of the society's highest honor, and, more pressingly, to his growing power and the threat that this posed to the future of the RAS. Faced with three factions to choose from—Lockyer, Proctor, or Airy, who utterly disdained both—the council, either through commitment to the Greenwich-Cambridge axis or through sheer disarray, failed to support either nomination and ultimately left the Gold Medal unawarded.

Within months matters would come to a head, with Strange using a private circular at the society's 1873 annual general meeting to try to oust Proctor from the council. The *Astronomical Register*'s detailed coverage of the ensuing argument describes "a long and stormy discussion" that included "conduct . . . hardly creditable to the oldest scientific society." A flurry of vitriolic letters from both sides followed in the journal, and although Proctor survived the coup attempt, within a year the fallout from his transit broadsides against Airy and his misuse of editorial privilege saw him too eased off the council.[26] Pritchard wrote to Lockyer to celebrate: "I hope you have had a sight of the Resolution inserted on the minutes of the R.A.S. regarding the abominable conduct of our late friend the Editor [of *Monthly Notices*]. . . . He must be at least a monomaniac—but his rabidness in future will be less mischievous."[27]

These matters of disciplinary politics illustrate very clearly the crisis of disciplinary identity and control that gripped English astronomy's governing elites in the wake of the introduction of astrophysical techniques in the 1860s. A massively increased plurality of viable practices opened up a greatly expanded scope for not just what counted as astronomical work but also for who might count as an astronomer and how they might present such labor. This created a wealth of challenges and open questions for the discipline's establishment elite concerning the control of authority, data, money, publications, public news, and institutionalized power. This crisis was at least a *three*-sided affair, pitching the Greenwich-Cambridge axis power elite of positional astronomy into a fight with both Lockyer's cohort of young, scientific naturalist, experimentalist professionalizers, and Proctor's "third way," disdainful of both the routine and unoriginal drudge work conducted at Greenwich and what he saw as the greedy, arrogant, exclusionary tactics of the X Club crowd. If Proctor's place in this disciplinary landscape has been almost entirely overlooked, then Airy's letter to John Couch Adams serves to illustrate how significant a threat he was deemed to pose at the time. Proctor was seen as a viable candidate to take control of the RAS, and it is evident that his work, critiques, and political machinations were a substantial factor in his discipline's crisis. This raises several important questions: How did Proctor attain so much influence so quickly? Why was he so disdainful of both rival factions in his discipline? And what, precisely, constituted his evidently powerful alternative model for correct astronomical practice? Pursuing the answers to these questions takes us to two intimately related topics: Mars and mass media.

MARS AND IMAGINATIVE ASTRONOMY

Victorian astronomy was news. Even when they weren't openly fighting, astronomers produced data, made discoveries, and presented novel claims that appealed to a large and rapidly expanding reading public. As recent scholarship has shown, this was a productive and symbiotic relationship. Media, rather than being parasitic on or subsequent to scientific work itself, fundamentally contributed to the working practices of these sciences.[28] Central to this claim is the argument that the genres through which practitioners communicate and the audiences with whom they engage must be understood as co-constitutive of knowledge itself. The ways in which genre affects the labor of authors, editors, and publishers, how it shapes the horizon of expectations of writers and readers, how it relates to the marketplaces that such work circulates within, and how audiences respond to such genre-bound works all shaped how Victorian astronomy worked.[29]

Lockyer and Proctor are particularly revealing actors in this regard. Both arose from outside traditional models of disciplinary training, both promoted novel approaches to the practices of astronomy, and neither had stable institutional bases from which to work. This meant that both men were forced to pursue other sites for representation, dissemination, authority, and approval. For both, their workplaces were profoundly and inescapably public. That the two clashed so violently is reflective of significant differences in how each man went about establishing his identity as an authority within such a cultural-scientific marketplace.[30] Lockyer's story is better known. Unhappily employed as a War Office civil servant, Lockyer's early career as an astrospectroscopist was frustrated by a lack of money and institutional support.[31] He soon threw in his lot with a collection of similarly discontented and equally ambitious young scientists, organized informally from 1864 around the X Club dining fraternity. Important for Lockyer's early assimilation into this milieu was his appointment as a regular scientific contributor to—and later shareholder of—the X Club–owned journal the *Reader*. Produced in furtherance of the club's ambitions to reform and modernize British science, the journal bucked populist trends in scientific journalism by rejecting the open, participatory, and generalist style of contemporaneous organs like the *English Mechanic and World of Science*.[32] The *Reader*'s owners aspired to establish a new, salaried scientific elite, and so the journal's stated goal was to "serve men of culture," stewarded by the self-styled "leaders of science in London."[33] Though it soon folded, its tone caught the eye of like-minded publisher Alexander Macmillan, who immediately bankrolled Lockyer in establishing his own journal, something of a successor to the *Reader*, called *Nature*.[34] Established in 1869, one of the first things Lockyer did with his new journal was to get into a fight with Richard Proctor.

Proctor's career up to the late 1860s had been one of struggle and perseverance, of, as he put it, "very uphill work." Seeking remuneration through writing had proved difficult, not least because serious works that bolstered his reputation as a research scientist, like the dense monograph *Saturn and Its System*, sold poorly. Articles in general interest and popular journals—often published anonymously—paid better, and after his financial ruin in 1866 Proctor began to split his time between works that generated income and his time-consuming research work into star distribution and drift that was intended for the weekly meetings of the RAS.[35]

By Proctor's own admission it was his turn to the subject of plurality—to the idea that the universe might be teeming with life—that began to garner him a large and loyal audience, enough to turn the heads of editors and publishers in London and to begin bringing in a reliable income.[36] Mars, in particular, fascinated readers as a potential abode of extraterrestrial life, and Proctor, with typical self-confidence, set about completely remapping the planet in 1867 (see figure 1.1). Mars was a tantalizing object at the very limits of telescopic scrutiny, inviting a heady mix of visual observation and analogical extrapolation. As K. Maria D. Lane has shown, Proctor skillfully deployed the perceived objectivity and authority of the map to make a case for the strong similarity between Earth and Mars.[37] This was not itself a particularly novel take; as both Proctor and his audience knew very well, the use of analogy in debates over the plurality of worlds had an extremely long history. For centuries the subject had pushed at the limits of inductive reasoning, forcing scholars from Immanuel Kant through Bernard Le Bovier de Fontenelle, Thomas Chalmers, David Brewster, and William Whewell to argue over the propriety of moving from the known to the unknown via analogical reasoning—or imagining—in order to conceive of worlds elsewhere.[38] Proctor, as Bernard Lightman has shown, rapidly took up the mantle of the English-language champion for a thoroughly modern defense of analogy, combining the findings of new astronomy with a Brewsterian belief in divine design to present a rousingly popular "cosmic, post-Darwinian theology of nature." On the one hand, such analogies relied upon a visual reading of Mars as inherently Earthlike in its appearance and geographical makeup. As Lightman notes, Proctor's new map "constituted the visual centerpiece for [his] proof that Mars could sustain life." Beyond maps, globes and stereographic photographs were also deployed to press home the case for Mars as "the Miniature of our Earth."[39] When the scientific

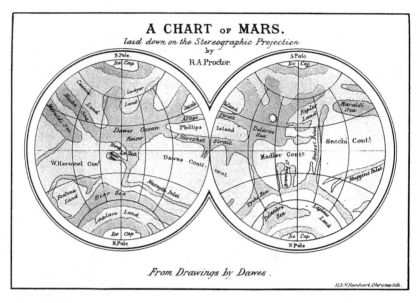

Figure 1.1. "A chart of Mars, laid down on the stereographic projection," originally composed by Proctor in 1867 using the observational sketches of the venerable Buckinghamshire astronomer William Rutter Dawes and included as a plate in his *Other Worlds Than Ours*.

instrument-maker John Browning constructed and photographed a new globe of the planet in 1869, Proctor wrote the explanatory text that accompanied the published stereograms, furthering his case for Mars as Earth's twin: "Throughout the whole of the solar system there is no object which affords so many points of resemblance to our own earth, or exhibits so clearly the signs of adaptation to the wants of living creatures, as the small orb which is our nearest neighbour among the exterior planets. . . . Mars presents a scene which requires no very lively exercise of the imagination to dot with villages, towns, and cities, peopled with busy workers."[40]

This account rested on more than visual similarities, however. Proctor substantiated and amplified these visual analogies with an even more powerful technique: physical analogy. With the new findings of the spectroscope to hand, Proctor could extend what was known about Earth's very soil and sky to what we might know about its outer neighbor. In *Other Worlds Than Ours* (1870), his first book-length treatment of the subject of plurality, Proctor skillfully linked evolutionary theories of progress with ideas of divine design, connecting both with spectroscopic evidence to transform celestial objects into living worlds elsewhere. "The mind," Proctor wrote, "is immediately led to speculate on the uses which those elements are intended to subserve." If iron, for example, was present on a planet such as Mars, "we speculate not unreasonably respecting the existence on that orb—either now or in the past, or at some future time—of beings capable of applying that metal to the useful purposes which man makes it subserve."[41] Only with the latest astrophysical evidence in hand could this crucial leap to physical use and purpose be made. From there, Proctor judged, speculation as to the analogous natural (theological) order of these bodies became only reasonable, and the only reasonable analogy was of a similarly living world, capable of exploiting the same materials man had already mastered. The book was, to say the least, a blockbuster, rapidly running through multiple editions and giving Proctor what he called his "long wanted for beginning." It would set the template for what would soon become a spectacularly successful career.[42]

Other Worlds Than Ours was a new kind of work, exploiting a surging genre—the cheap popular science book—to present not just a new argument for the plurality of worlds but also a new model of astronomical practice. I suggest that this work and Proctor's subsequent work in this style be considered under the umbrella term "imaginative astronomy," not to highlight any unique claims on Proctor's part, but rather, on the one hand, to focus attention on the fact that this cluster of methods needs to be recognized as a form of astronomical practice (rather than mere popularization) and, on the other hand, to highlight the historical specificity of what made these particular methods successful—and controversial—when and where they did. By coupling analogy so directly with the cutting-edge work of the new astronomy, Proctor made a double move: he placed his readers within an imagined environment, inviting them to extrapolate their own experiences and understandings to consider the lived environment of other worlds. And he substantiated these profound speculations with an imaginative, deductive form of reasoning that marshaled a suite of state-of-the-art scientific evidence to surmise the physical state of those worlds elsewhere.

Such a strategy was highly controversial, not least because it challenged what William Huggins had called the "strictly fiducial" role of terrestrial-celestial spectral comparison. For Huggins, Lockyer, and their allies, the maintenance of the fragile validity of their striking extension of laboratory trials into the heavens depended upon the establishment of the spectroscope as a highly accurate instrument, licensed by the precision standards of the Maxwellian laboratory.[43] To its critics, therefore, *Other Worlds Than Ours* attacked the very foundations of this empirical metrology. An anonymous reviewer in the conservative *Saturday Review* summed up the division clearly. Proctor's book, they lamented, had reduced *inductive* proof to "the subordinate part of putting a veto upon what the speculations of fancy or the bent of popular persuasion might see fit to propound." This was "carrying the war with a vengeance into the country of such as stipulate for proofs of a positive kind. They are to be saddled with the proverbial difficulty of establishing a negative." After all that had been done in recent years "to lay the groundwork of philosophy on the accumulation of facts aided by the logic of induction," they asked, "are we to see a school in science prepared to lay its foundations and superstructure alike upon what were fondly thought the exploded principles of teleology?"[44]

Responses like this to Proctor's book formed part of a wider contest over the role of science in English society in the early 1870s. This was a particularly unstable period for metropolitan elites, both old and aspiring. The younger generation's challenges to the prerogative of Victorian gentlemen of science constituted part of a larger crisis of authority for aristocratic elites and their church base, proliferating a complex array of contests for cultural authority.[45] Theodore Porter has cogently argued that the rise of scientific naturalism had less to do with an urge for professionalization than it did a desire to reconfigure science's cultural authority in the face of new threats from working-class radicalism and novel modes of popular readership and expression. As they made gains against those above them, the scientific naturalists sensed a threat from below. So, like the gentlemen aristocrats they worked to replace, the scientific naturalists fought hard for their own prominent place in wider culture, such that they too could reshape it in their image. Their ideology, though seemingly radical in the heady days of the X Club's mid-1860s rise, soon shifted toward a new formulation of respectable science in the public sphere. Threatened by radical and speculative forms of populist science, the scientific naturalists promoted a swath of modernizations that were broadly characterized by rational, naturalistic, secular discourse, produced and policed by a new elite certified by institutional qualifications and renewed specialist societies.[46]

The role of mass media was both a central problematic and a potent weapon in this battle. *Nature*, as historians have recently shown, epitomized the approach forged by Lockyer's modernizing cohort. Though Lockyer had initially promoted it as a general interest periodical suitable for the educated layman, his journal soon fell in with the broader forces of scientific naturalism. Within a few years of its founding, *Nature* had become a small-circulation niche journal catering to the needs of young men of science keen for a shared forum for fast, independent, cross-disciplinary communication and debate.[47] Implicitly at least, its aim quickly became to serve as a locus of control and critique in wider cultural discussions about science. Achieving this goal required establishing a particular kind of authority as a scientific journal, certainly not a given in the flux of the early 1870s. Telling, then, is the Royal Society's novel and monumental effort to construct a supposedly comprehensive *Catalogue of Scientific Papers*. This was, it has recently been argued, ultimately a project to establish the privileged status of the named-author scientific article at the expense of books, newspapers, and general periodicals. It was also, therefore, a project aimed at demarcating the boundaries of what did and did not count as a legitimate venue for such *memoirs*. In selecting journals appropriate for the catalogue, "social and scientific respectability" became an abiding concern of the society's elite decision-makers. *Nature* made the cut. The sorts of platforms that Proctor wrote for most certainly did not.[48]

Other Worlds Than Ours was not just a bad book, then—it was a threat. Imaginative astronomy challenged the scientific naturalists' model for cultural authority at a moment when that authority hung in the balance. Lockyer had to respond, so he turned to his friend Charles Pritchard to review Proctor's fast-selling book. The Savilian Professor of Astronomy at Oxford, Pritchard was an astute choice. As both a grandee of the RAS and a subtle modernizer with close ties to the Lockyer crowd's ambitions for astrophysical research, he straddled both camps aligned against Proctor. Writing to Lockyer to accept the job, Pritchard was frank about his distaste for Proctor's "uncomfortably written" book. In private, this was a judgment explicitly linked to Proctor's peculiar position as a worker committed to supporting his scientific labor through popular writing. "Something's the matter," Pritchard confided to Lockyer; "I fear he has a hard life-struggle and unites too much."[49] Translated to the bounded semipublic arena of a *Nature* book review, Pritchard turned these concerns into a moral lesson on the right and wrong ways to be an astronomer. Proctor had criticized some of Lockyer's work in his book, and Pritchard fired back at Proctor's impropriety in straying outside the bounds of appropriate disciplinary decorum. Audiences mattered a great deal in this ethical lecture. Having failed "to settle in his own mind what particular class-es of persons the work was intended to suit," Pritchard wrote, Proctor had erred in applying spectroscopic evidence to the "not very fruitful question of the habitability of the planets," resulting in a style that sometimes "border[ed] on dogmatism."[50] On this telling, it was precise-ly the link between novel spectroscopic data and pluralist speculation that discounted Proctor's claim to astronomical authority.

Proctor, in an angry letter of response, charged Pritchard with misquoting his book and claimed that the review was "vitiated by mis-statements or omissions, which one can scarcely imagine to result from mere negligence."[51] Lockyer happily printed this riposte in *Nature*, not least because it afforded him the opportunity to append a stinging ed-itorial footnote beneath it. "Astronomers did not require Mr. Proctor to tell them what he has recently been enforcing," he chided his col-league, "but, more modest than he, they have been waiting for facts." Surely, Lockyer wrote, "Mr. Proctor . . . is old enough to see that by attempting to evolve the secrets of the universe, about which the work-ers speak doubtfully, out of the depths of his moral consciousness, he simply makes himself ridiculous, and spoils much of the good work he is doing in popularising the science."[52]

By flagrantly exploiting his role as editor, Lockyer seemed to have gotten the last word on the matter. Most damningly, he had used his chiding footnote to throw into stark contrast the role of astronomers, as "modest . . . workers," with the role of "popularizers," ridiculous in their flights of fancy and juvenile in their devotion to a cheap-book-buying mass audience. Proctor's offense at being branded no worker was no doubt amplified by the proximity of Lockyer's practices to his own. Lockyer's precarious position as a civil servant and editor made him, too, dependent upon the unstable array of income sources linked to popular astronomy: lecture fees, journalism, and the largess of a pub-lishing house ostensibly keen to make science publications pay.[53] Genre was therefore a crucial tool and marker for distinguishing between the practices of the two. Proctor had already been quite explicit about his disdain for the way Lockyer conducted *Nature*. A week before the appearance of Pritchard's review, he had openly derided it as "about the worst edited journal I know of."[54] This attack had been launched, quite pointedly, from Proctor's own favorite location for comment and debate, a popular journal that had most certainly not made it in to the

Royal Society's *Catalogue of Scientific Papers*: *the English Mechanic*. It was here too that Proctor chose to move the fight over *Other Worlds Than Ours*.

Typically contrasted with *Nature*, the *English Mechanic* (founded in 1865) was one of the few truly successful science magazines to emerge from the periodicals boom of the 1860s. By the early 1870s, the journal had established for itself a large and loyal working- and middle-class readership, many of whom were active participants in its communal, collaborative scientific and technological discourse.[55] Proctor lauded the publication for its egalitarian approach and effusively praising the skill with which the magazine "managed [its] Correspondence columns." From 1869 onwards, he became both an author of articles for the journal as well as a leading contributor to its letters pages. As a result, he would already have been very well known to most of the journal's readers by the time of Pritchard's review, which stood in stark contrast to the *English Mechanic*'s judgment that *Other Worlds Than Ours* was "thoughtful, suggestive and very interesting."[56] As allies and reciprocal advocates, Proctor and his readers can be situated in a particular social space then in the making. The *Mechanic*'s promotion of pugnacious and participatory scientific debate enabled Proctor to fashion for himself an identity as a particular kind of public expert. His regular "Column of Paradoxes" for the journal viciously savaged men like the flat-earther John Hampden, establishing Proctor's own bona fides and demarcating the boundaries that distinguished him from the failures of institutional astronomy.[57] By highlighting the inability of traditional scientific elites to properly police their disciplines, Proctor positioned himself as a crusader on behalf of just the sort of self-improving scientific generalists that constituted the majority of the *Mechanic*'s readership. In the words of his friend Edward Clodd, Proctor was a man who "hated shams and quacks, of whatever profession." To his devoted readers, therefore, Proctor was precisely what an expert looked like. "It is rather too bad," wrote one *English Mechanic* correspondent, "to expect such men as Mr. Proctor to waste time and trouble about such matters; it is like getting a professor of six or seven languages to argue with a precocious juvenile."[58]

From such a position of strength, Proctor could confidently deride the *English Mechanic*'s much less successful competitor, *Nature*, as a "contemporary of minute circulation." (On these terms alone, it was not a fair fight. The *English Mechanic* had a circulation of somewhere between thirty thousand to eighty thousand, whereas the loss-making

Nature struggled with subscription numbers in the hundreds). Proctor's response to Pritchard's review continued to pull no punches. The review was, he wrote, "more discreditable . . . than any which the veriest hack-critic would have penned," and he proceeded to dismantle it almost line by line.[59] In a long and distraught letter to Lockyer, Pritchard complained that he had "been exposed to an avalanche of abuse and vulgarity from Mr. Proctor" and that it was not "becoming" of Proctor to complain about his review in another publication while the pages of *Nature* were still open to him. In urging Lockyer to let the matter drop, Pritchard summed up the gulf in attitudes toward astronomical practice that existed between the two battling camps: "Mr. P. evidently has a motive for keeping himself loudly before the public. But he will hang himself with enough of rope. . . . What grieves me sadly is this—astronomy hitherto has been singularly free from noise and controversy. Chiefly because our truths were evolved not from our 'consciousness,' but from observation and geometry." By replying to Proctor, Pritchard warned, "one only plays his own game, viz into his notoriety."[60]

These responses made in private and in public by men like Airy, Lockyer, and Pritchard clearly link a distaste for speculation and imagination with a characterization of such work as merely—or notoriously—"popular." Historical epistemologists have noted how Enlightenment projects that aspired to the universal communicability of the sciences begat crises over the place of imagination in scientific thought. The emergence after 1860 of new and diverse audiences for astronomy evidently made trouble for the supposed mechanical objectivity that Pritchard so desperately pined for.[61]

IMAGINATION AND THE MORAL PURPOSE OF ASTRONOMY

Proctor, of course, very much recognized the pejorative sense in which his critics dubbed his work as "popular," and he did not shy away from developing a forceful rejection of this characterization. In the wake of the fallout over *Other Worlds Than Ours*, Proctor doubled down on his imaginative approach, and in the years following he would come to articulate a very clear defense of it. This defense was built around the question of astronomy's *purpose*. Like Pritchard's ideal "observation and geometry," this was a moral as well as a practical matter. While his opponents staked their claim to cultural authority on the exacting precision of ordered, regulated, and cloistered labor, Proctor set the goals and ambitions of his discipline higher. Astronomy, he came to assert, was fundamentally and irreducibly a science not of measurement but of *enlightenment*. Its ultimate purpose, therefore, was nothing less than as a "potent means of culture." The astronomer's job, in his view, was to clearly and persuasively explicate the wonders and majesty of the universe to all who cared to read and listen. Successfully executed, this work would achieve nothing less than the raising of the "mental and moral culture" of society at large.[62] In contradistinction to his opponents' fears of runaway imagination in the service of rank populism, Proctor countered with a cogent and powerful moral case for a necessarily public form of astronomical labor.

Such an idea was by no means sui generis. Some decades earlier, John Herschel, in his hugely popular 1833 *Treatise on Astronomy* (and again in its 1849 revision as *Outlines of Astronomy*), had described the pursuit of scientific reasoning as "the first movement of approach towards that state of mental purity which alone can fit us for a full and steady perception of moral beauty as well as physical adaptation."[63] Motivated by this moral imperative, Herschel had gone on to write of the stars as "effulgent centres of life and light to myriads of unseen worlds," and he had delighted in presenting dramatic and imaginative descriptions of how the heavens might appear to inhabitants of planets orbiting our sun and even other stars.[64] If such flighty evocations had become controversial several decades later, then the occasion of Herschel's death in May 1871 afforded Proctor an ideal opportunity to remind both his readers and his enemies that the century's greatest astronomer had also been one of its most imaginative. In three extended obituaries, Proctor used the lessons of Herschel's life to make clear for his readers the fundamental distinction between those who merely observed and those who shaped observations into astronomical understanding. The former, as Proctor characterized it, was merely a "practical" art, typified by the rote data gathering at a place like Greenwich, and conducted toward goals no more elevated than the regulation of timekeeping and the safe navigation of ships. The latter, in contrast, as typified by the towering achievements of the recently deceased Herschel, was wide-ranging, speculative, and synthetic, taking as its goal nothing less than an understanding of the universe. Its purpose, therefore, was as noble as it was simple: "the extension of our *knowledge*." On this telling, Herschel's life was a vital lesson in disciplinary achievement. Astronomy's highest rank, Proctor noted, belonged only to those who, like John and his father William, had advanced astronomy not through mere mechanical observation but rather through "judicious theorising." Norman Lockyer's editorial attack could, as a result, be neatly turned back against him. "Observers," Proctor wrote, "at least such observers as do not themselves care to theorize—are apt to contemn the theorist, *to suppose that the hypotheses he deals with have been evolved from the depths of his moral consciousness,* instead of being based on those very observations which they mistakenly imagine that the theorist undervalues." Neatly inverting the presumed hierarchy of the Greenwich-Cambridge axis, Proctor concluded that it would be "as reasonable for the miner to despise the smith and the engineer, as for the observer in science to contemn him who interprets observations and educes their true value."[65]

In private correspondence, Proctor had already secured from his hero what he took to be a personal endorsement of these views. Proctor had written to Herschel shortly before his death, both to discuss his latest theories on the structure of the universe and to moan about the state of their discipline. "Many view the careful study and analysis of observations, work which they call 'paper astronomy,' with contempt," he complained. "I should like to see many of those now engaged in accumulating preposterously useless observations at work in the field over the fence of which I have looked."[66] Herschel's reply—which Proctor ostentatiously published after his death—offered guarded support for this model of astronomical theorizing. He certainly felt no "contempt for astronomy on paper," he told Proctor, and he hoped that "you will not be deterred from dwelling more consecutively and closely on these speculative views by any idea of their hopelessness which the objectors against paper astronomy" may have offered. "*Hypotheses fingo* in this style of our knowledge is quite as good a motto as Newton's *Non fingo*—provided always they be not hypotheses as to modes of physical action for which experience gives no warrant."[67] Proctor's various and extensive eulogies to Herschel therefore served to promote this idea of *hypothesis fingo*, while at the same time setting Proctor himself up as the natural new bearer of this particular astronomical torch. Astronomy should be studied and practiced, Proctor asserted, "*as a means of mental training,* whether as affording subjects of profitable contemplation, or as offering problems the enquiry into which cannot fail to discipline the mind."[68]

Proctor, however, faced a very different disciplinary landscape in the early 1870s than Herschel had four decades earlier. Where Herschel wrote as astronomy's leading voice, Proctor wrote as an upstart agitator at a time of great uncertainty for such a publicly oriented approach. That mental and moral improvement through astronomy was and should be accessible to everyone clearly motivated Herschel's seminal popular works, but his successors in astronomical research were evidently not so sure. Proctor was not shy in pointing out how directly this came down to a falling out over styles of exposition. "In considering astronomy as a subject of study," he wrote in one of his Herschel obituaries, "the first point to which we must direct our attention is the mode in which astronomical discoveries should be presented."[69] Because this was a question of genre, such disagreements tangled authorship and readership up with that most vexing of late Victorian issues, the funding of science. If science's principal purpose was the raising of society's "mental and moral culture," then popular authorship became, as far as Proctor was concerned, the ultimate arbiter of a practitioner's competence. Market responses were, in effect, the only relevant proof of an astronomer's ability. In 1876 Proctor codified this doctrine into a strident manifesto, *Wages and Wants of Science-Workers*, arguing forcefully against state handouts and for the necessary link between good science and the public's willingness to pay for it. The "most fruitful of our *scientific workers*," he wrote—Herschel, Charles Lyell, Charles Darwin, John Tyndall, Alfred Russel Wallace—"are also those who have succeeded best in scientific literature." Again, hierarchies— disciplinary, economic, and moral—were inverted. While "*soi-disants* professional astronomers" survived only through the "jobbery" of state handouts, Proctor's genuine "science workers" flourished within a virtuous circle of commercially viable authorship and creative, engaging research.[70] The particular appellation of "science worker," then, must be recognized as another significant element of Proctor's identity formation. Just as the "man of science" was an identity forged through a complex set of discriminations intended to exclude other groups from the making of scientific knowledge, so Proctor's self-identity as a "science worker" was one vital component of his rival conception of a productive, democratic, anti-elitist, publications-based, socially embedded *profession.*[71]

Proctor's case therefore throws into sharper relief the fraught relationship of imagination with objectivity in this era. John Tyndall's well-known (but tentative) endorsement of imagination before the British Association for the Advancement of Science (BAAS) in 1870 and his success as a popular lecturer and author had made him a science worker, in Proctor's eyes.[72] Yet the role of imagination in aiding or hindering the crucial move from data to principles and theories remained the central question for arguments over right practice across private and public realms. Within three years of Tyndall's 1870 address, P. G. Tait had launched a vicious attack on the Royal Institution professor, accusing him of having "martyred his scientific authority by deservedly winning distinction in the popular field."[73] Tellingly, by 1878 Tyndall had shifted position: "It is against [the] objective rendering of the emotions—this thrusting into the region of fact and positive knowledge, of conceptions essentially ideal and poetic—that science, consciously or unconsciously, wages war."[74] Proctor saw exactly the opposite lesson. "No one," he asserted in his provocatively titled *The Poetry of Astronomy*, "who studies aright the teachings of the profoundest students of nature will fail to perceive that [they] have been moved in no small degree by poetic instincts, and that their best scientific work has owed as much to their imagination as to their reasoning and perceptive faculties."[75] The nature of expertise and of the expert loomed large in these contests. Doing astronomy imaginatively, Proctor was keen to stress, absolutely did not diminish the demand for expertise in its practitioner. No one, Proctor wrote, can "write really useful popular treatises on science who has not himself undertaken scientific researches. . . . In such cases the scientific habit of mind ensures accuracy in the mode of treatment." Expertise alone was not enough, however, if one's work came across as "unimpressive" or "wearisome." In order to successfully work the "raw materials" of astronomy into "the manufactured article—Knowledge," the astronomer must move beyond dry accounts of gathered data and begin to speculate on the causes and meanings behind them.[76] To presume that Tyndall and his allies' shift against such imaginative practice left Proctor on the margins is both to miss Proctor's not-insignificant influence and power and to presuppose the ultimate success of scientific naturalism. Alternative models were in play, and they agitated much of the fear that animated men like Lockyer.

The role of commercial publication was a central battleground in these contests. Proctor advanced his own arguments in an extraordinarily wide range of scientific and general interest periodicals, newspapers, books, and extremely successful lectures. At the same time, he castigated colleagues who used or promoted publications controlled by or exclusively aimed at the soi-disant professional astronomers. When Airy fretted about Proctor leading the Royal Astronomical Society to the dogs, it was pretty clear to men like him who those dogs were.

Audiences mattered because the expectations of readers and listeners shaped how novel claims at the forefront of astronomical research were formulated, synthesized, communicated, and interpreted. Simon Newcomb, America's foremost positional astronomer and very much the model of a conservative practitioner, expressed dismay in a review of Proctor's *Borderland of Science* that the author had presented speculative material, including a chapter on "Life in Mars," as "scientific truth." In a cutting reminder of his faction's expected norms of practice, Newcomb urged Proctor to "leave this 'border-land' between science and nonsense to be cultivated by hands capable of nothing better."[77]

These were concerns that clearly bridged the roles of author and astronomer. Newcomb himself wrote prolifically for a mass audience, but he worked with a very different ideological toolkit when it came to matters of appropriate forms of discourse. In reviewing Proctor's recently completed "great run" of lectures in the United States in 1874, Newcomb judiciously waited until Proctor was on the boat back home before setting forth his opinion of the English astronomer's tour:

> Our judgement of his lectures must depend on the standpoint we take. His hearers were pleasantly entertained, saw multitudes of pictures of comets, satellites, and nebulae, and went away with many new ideas of the interior constitution of Jupiter, the protuberances of the sun, and the probable inhabitants of other worlds. The questions, whether such nebulae and star clusters were ever really seen with a telescope . . . and whether there is any trustworthy evidence in favor of Mr. Proctor's opinion that Jupiter is white-hot, is one which some may deem relatively unimportant. Taking this view, the lectures were a great success. But if we take the ground that a popular lecture should give a clear, correct, and connected view of the past and present state of the science with which he is dealing, then Mr. Proctor is less worthy of praise. He is too incautious in his statements, too rash in his speculations, and too diffuse in his mode of treatment.[78]

Newcomb makes a clear distinction here between visual representation and mathematical evidence. His review contrasts what is merely seen with what is grounded in calculation, a critique that implicitly demarcates and delimits the role of public audiences. An audience's horizon of expectations, its desire to be entertained as well as educated, its preference for visual spectacle over mathematical rigor, and its ability to define the marketplace for such imaginative astronomy greatly problematized and therefore actively shaped the conflicts that arose out of astronomy's growing plurality.[79] This gave practitioners a wider scope for self-fashioning as "scientific experts" than Newcomb wanted or the RAS could possibly control. Newcomb's argument therefore ran both ways, as Proctor knew better than most. Responding to a review ridiculing the imaginative tone and spectacular frequency of his books, Proctor gave as clear an answer as he could: "as respects *the progress of science* and the spread of scientific knowledge, I should have thought a popular book read by thousands must be more useful than a more profound book read only by a few students."[80] In Proctor's eyes, good science was *by necessity* popular. Public engagement drove scientific progress, just as scientific progress, properly explained, drove the development of society's mental and moral culture. The resources that made this ideology work came from new journalism.

NEW JOURNALISM AND THE MARKETPLACE FOR IMAGINATIVE ASTRONOMY

At this point, it's worth pausing to reiterate the specificity of the arguments presented here. Richard Proctor's 1882 letter to the *Times* regarding the canals on Mars can only be understood, I am suggesting, once its particular contexts of production and reception have been recovered. It is no help at all to think of this letter, or any of Proctor's works, as merely part of "Victorian popular culture," given the dramatically different working worlds of public astronomers such as John Herschel and those such as Proctor who worked several decades later. Even within the overlapping public spheres in which Lockyer and Proctor both practiced, we have seen how dramatically different methods and audience responses emerged and developed as part of contingent and conflicting models for correct astronomical practice. So if Proctor's moral project of public betterment looks superficially similar to other stories from the history of science—for example John Pringle Nichol's advocacy of a reformist agenda through a "science of progress," or the educational program of the petty-bourgeois Edinburgh Philosophical Association, or the diffusionist ambitions of the X Club's International Scientific book series—we must remember that each of these programs developed and was shaped in response to sociohistorically specific circumstances.[81]

Unique to Proctor's moment, I want to suggest, is a particular cluster of resources, constraints, and material products that together defined a marketplace for imaginative astronomy. If the first half of this chapter has explicated what imaginative astronomy was, this second half aims to explain why it took the form that it did. Proctor's transatlantic success as an author, editor, and astronomer can be linked to his growth from and exploitation of new developments in Anglo-American journalism that began in the 1860s and reached their peak in the early 1880s. Proctor's program for an open-minded, democratic, and anti-elitist scientific practice was an emergent product of this moment, and its success hinged on changes in Victorian society that produced both practitioners and an audience mutually committed to these ideals. We can see the essence of this new mood in, for example, a review by the liberal *Manchester Guardian* in 1874, assessing recent works by Proctor, Tyndall, and Robert Kalley Miller:

> The intercourse between the purely scientific and the popular mind is daily becoming closer. Conservatism in science is an impossibility, and the days are gone forever when savans wrote in barbarous Latin that they might screen their lucubrations from vulgar eyes. So intimately have scientific questions become associated with modern opinions on anything worth thinking about, that no intelligent man can afford to be ignorant of them. The best proof of the increased popular interest in scientific enquiry of every kind is to be found in the number of works and editions that issue from the Press. Physical as well as natural science has not only widened its sphere of discovery, but largely increased the number of its disciples.[82]

The marketplace central to this moment came to be known (much later) as the new journalism. As an upshot of growing and diversifying mass media, the new journalism can trace its roots to now well-recognized transformations in British mass media in the second half of the century. The gradual repeal of the "taxes on knowledge"—including advertising duty in 1853, newspaper stamp duty in 1855, and paper duty in 1861—and liberal education reform both served to greatly expand the reading public. A slew of new technological developments—the telegraph, typewriter, telephone, wood pulp paper, high-speed rotary press, and half-tone block for the reproduction of photographs all came into regular use in the press industry between 1860 and 1900—greatly expanded the capacity of the media industry to service this new marketplace.[83] New techniques of production and editorial control were integral to this rise, dependent as it was on securing the loyalty of new and discriminating readers. The rise of "class and trade" journalism that targeted specific markets was one upshot of these transformations, and the development of new norms of practice in daily newspapers was another. Though commonly dated to the 1880s, recent scholarship makes a persuasive case for the gradual emergence, starting in the 1860s, of the democratic and anti-elitist journalistic practices that would come to define Britain's new journalism. The traditional idea of a rupture between "old" and "new" forms of journalism has been replaced by a growing consensus that these changes unfolded progressively. Their origins are now traced back to both the gradual incorporation of editorial techniques developed first in the United States, as well as the same taxation and regulatory reforms that gave rise to the proliferation in "class and trade" journals.[84]

These changes both fed off and reflected significant developments in British society. The 1860s—what Eric Hobsbawm called the "decade of reform"—saw rapid economic growth and liberalization coupled with increased social mobility and greater political agency for the middle and working classes. The impact of Henry Mayhew's groundbreaking *London Labour and the London Poor* (1851, 1861–1862 and 1864), the *Lancet*'s criticism of workhouse conditions (1865–1866), the socialists' First International (1864–1876), the success of Samuel Smiles's *Self-Help* (1859 and 1866), the expansion of mechanics' institutes as sites of working-class nontechnical education, the growth in strength of the Women's Movement, the Reform Act of 1867, and the Education Act of 1870 are all indicative of these trends.[85]

Journalism's power to influence and reflect this changing political mood has been linked to its transition toward the deployment of a market-oriented "objective gaze." After the mid-1850s, the highly politicized rhetoric of the "publicists" who controlled the radical early Victorian unstamped press gave way to a new, specialized, and increasingly autonomous field of discursive production based on a disinterested form of reportage. As direct forms of media control were removed, the dependence of the press on political patronage or radical underground backing gave way not to press freedom but rather to new forms of market-driven capitalist control, placing new demands on editors and journalists to produce copy that would sell. Print became cheaper and its markets correspondingly expanded, industrializing the press and effectively forcing most papers and journals to pursue high circulations and advertising revenue. Increasingly, this meant the marginalization of the radical press and the replacement of political grandstanding and firebrand rhetoric with a form of "news"-based journalism intended to appeal to a broad and mostly middle or skilled working-class audience.[86] Though much of this output remained political in character, its tone shifted to one that would ensure broad appeal. The result was an upswing in coverage of social ills and criticisms of political privilege, content that focused on liberal ideas of reform popular among the consumer classes, rather than calls for revolution or attacks on capitalism that were not.[87] A tone of objectivity better fit this liberal appeal to gradual reform through education and cooperative betterment, ideas that relied heavily upon popular emphasis on individual knowledge-acquisition as a path to self-improvement and economic progress. Increasingly, the press saw its role as providing this knowledge through reportage that sought to uncover facts rather than present opinion. The political stance of this new media was therefore genuinely populist, insomuch as it promoted the idea of social cohesion and moral improvement through the spread of middle-class enlightenment. This in turn made it cautiously anti-elitist, in the sense that it played an important role in promoting gradual reforms to the political structure. In particular, this media's democratic campaigning style was integral to mobilizing working-class support for political parties, helping convert them from "what had been essentially aristocratic factions in Parliament" into "mass political organizations."[88]

The press campaign as a technique of political agitation and readership acquisition grew out of these political and social transformations. From Charles Dickens's hugely successful "'uncommercial' philosophy" aimed at exposing the unjust to the *Pall Mall Gazette*'s early agitations over workhouse conditions and "baby farming," moral campaigns and an all-around adoption of a more idiomatic and familiar style of prose began to transform the press into what one later commentator called "the literary mirror of the talk, and therefore of the mind, of the average intelligent and educated Briton."[89] It was the marriage of this democratic and campaigning style with economic and editorial techniques copied from American metropolitan dailies that transformed the liberal investigative journalisms of the 1860s into the fully fledged new journalism that began to appear in the 1870s. Faced with fierce competition for the nascent and expanding markets of mass readership, a new generation of ambitious editors and journalists worked to make print culture more accessible and profitable by significantly altering conventional newspaper practices, marrying sensationalized news coverage and format and distribution innovations with a populist agenda of social reform.[90]

This new journalism came to be synonymous, above all, with

the strong, vocal, and campaigning editorial style epitomized by the genre's most enduring hero, William Thomas Stead. Between 1871 and 1880, Stead—then the youngest newspaper editor in the country—had transformed the Darlington *Northern Echo* from a provincial Liberal daily into a campaigning paper of national repute. This early success earned him a position as deputy editor of London's *Pall Mall Gazette* in 1880, replacing John Morley as editor in chief three years later. Morley himself never practiced much as an editor, leaving Stead from the beginning of his time at the *Pall Mall Gazette* to impose his signature style on the paper. As one later commentator observed, this style was epitomized by a "mingling of democracy and sensationalism" intended to both capture a wide readership and act as a force for political and moral betterment.[91]

Stead would come to play a significant role in Proctor's career, both as a personal champion of the astronomer and as a major influence on his editorial style. The parallels between the two men's epistemologies and editorial techniques are therefore worth exploring in some detail, as they speak of the shared cultural moment in which they found their voices and their overlapping audiences. Both secured their fame through strong, puissant, but trustworthy personas, built upon a prolific style of accessible writing that skillfully elided the roles of expert, author, narrator, and editor. (Critics of both pointed to the sometimes difficult task of separating hard fact from narrative creativity in their work). Stead, like Proctor, advocated a progressive liberal creed that relied on often vicious attacks on the inveterate power of social and political elites. Though commercially driven, both claimed a moral basis for their campaigns of public betterment, and both pushed for the opening up of public access to information and knowledge. Stead's now-famous manifesto for "Government by Journalism" proclaimed that the purpose of his profession was nothing less than the "lift[ing] of the minds of men . . . into a higher sphere of thought and action."[92] This self-styled "public duty" relied upon the marshaling of public opinion in order to coordinate a power base through whom one could claim to speak. Stead, like Proctor, worked assiduously to cultivate a unified and involved readership engaged through open discourse.[93]

Neither man should be seen as an isolated figure, however. These were not individual projects so much as strong articulations of a wider mood of social and political reform and a concomitant desire for the democratization of knowledge. Both were successful because of the public support they could marshal, and both worked alongside numerous allies committed to the same mass movement of public moral and intellectual improvement. Proctor's allies were strong enough, at their peak, to get him nominated for the RAS's Gold Medal, and his public support swelled to such an extent that there was agitation—in the press at least—for his appointment as Airy's successor as Astronomer Royal.[94] Stead too was part of a large network of allies and organizations upon whom his sensational campaigns were dependent, ranging from friends in high places (godfather of the Liberal Party William Gladstone, U.S. industrialist and peace campaigner Andrew Carnegie) to radical accomplices (socialist activist Annie Besant, women's rights campaigner Josephine Butler) to major popular organizations (the Salvation Army, the Society for the Protection of Children).[95]

The power of these emergent mass movements also provided much of the animating force behind attacks on both men. Faced with mass political engagement and broad demands on access to knowledge, Proctor's and Stead's critics perceived not only a loss of individual liberty (and with it, of course, elitist control) but also a troubling mingling of facts and sentiment within the suddenly influential media that grew out of and fed into these movements. Matthew Arnold's famous attack on what he contemptuously dubbed this "New Journalism" therefore sounds strikingly similar to Simon Newcomb's attack on Proctor: "The New Journalism . . . is full of ability, novelty, variety, sensation, sympathy, generous instincts; [but] its one great fault is that it is *feather-brained*. It throws out assertions at a venture because it wishes them or itself, if they are false; and to get at the state of things as they truly are seems to feel no concern whatever."[96]

At the root of this epistemic panic was a more fundamental fear about the nature of mass media in an age of transatlantic technology. An anonymous critic in the always conservative *Saturday Review* captured this angst at the same time as diagnosing its root cause: "The 'New Journalism' is cheap, and it is nasty, therefore it is popular. . . . Rational men need not buy and read the ill-printed, ill-written, ill-conceived rags of the cheap press. . . . When everybody can read, when the least educated and worst bred are the largest public, of course the wants of that public will be supplied, and even anticipated. It has long been so in America; and now that the *nidus* of the disease has been prepared in England, we have the malady."[97] This was the flip side to what Clare Pettitt has described as a reestablished Anglo-American alliance forged through the mass media in the wake of the laying of the first successful transatlantic submarine cable in 1866. This alliance, she notes, was "based on the 'new' values of the American Union—values of democracy and freedom."[98] As much as the *Saturday Review* might have reviled this movement, Stead and Proctor embraced it. The implication of Proctor as a new journalist becomes most clear, in fact, when he is placed in this context alongside Stead as a vehement advocate of the Americanization of British science and society.

At its most obvious level, this Americanization constituted the incorporation of a diverse range of journalistic innovations pioneered first in the urban centers of the United States. From the 1830s, commercial imperative and a "ferment of pre-literate popular culture" mingled with Jacksonian democracy to produce a proliferation of cheap American newsprint. It would be in this "penny press" that many of the norms of the later new journalism were first developed. By the 1870s, American editors like James Gordon Bennett Jr. at the *New York Herald* produced newspapers that sold in the hundreds of thousands, attracting a diverse readership with bold headlines, sensationalist stories, and popular campaigns.[99] These successful tactics of "yellow journalism" soon crossed the Atlantic. News-based journalism, one historian has argued, "is an Anglo-American invention." As one New Zealand observer commented in 1875: "the example of the *New York Herald* is now being largely followed by the chief London journals, and the great strides of modern journalism is one of the astonishing progressive features of a progressive age." The *Times* of London, somewhat less impressed, dubbed Stead's upstart *Pall Mall Gazette* "fourth-rate Transatlantic journalism."[100]

Underlying this story of transatlantic media exchange is a more fundamental exchange of political and social ideas. The new journalism's most significant import was not editorial, typographical, or journalistic; it was ideological. Proctor and Stead both explicitly identified the United States' popular zeal for knowledge and its ethos of individual self-improvement as the key lesson learned from their extensive travels there, and both men would establish these ideals as the new journalistic basis for their attacks on the elitism of old Europe. Stead, a longtime advocate of Anglo-Saxon race unity and Anglo-American unified citizenship, made his own views on the preeminence of American ways of thinking and doing quite clear, most dramatically in his rabidly pro–U.S. manifesto *The Americanization of the World, or The Trend of the Twentieth Century* (1901). For Stead and his allies such as Cecil Rhodes it was quite evident that a shift in the balance of intraracial power had left the United States ahead of Britain in most aspects of social and economic life.[101] Yet rather than take this fact as a warning or a threat, Stead argued that the preeminence of the United States should be both celebrated and embraced: "The creation of the Americans" was, he wrote, "the greatest achievement of our race," and the export of their values through Americanization would be the defining trend of the century to come. Stead himself had long been at the vanguard of this process, and his reflections upon the American press were at the same time a rallying cry for his own new journalism: "American journalism, as compared with that of Great Britain, is more enterprising, more energetic, more extravagant, and more unscrupulous." This last word was very much meant as a compliment. Though the English might sneer at U.S. newspapers, for Stead their virtues— their democratic spirit, their objectivity and honesty, their variety, their exuberance of design and coolness of style, and their liberty from political and social restrictions—put them "distinctly ahead" of their English contemporaries.[102]

Proctor likewise was a keen proponent of American cultural and intellectual norms and ideals. Through the course of his career he traveled the North American continent extensively on numerous lecture tours, eventually marrying an American and emigrating to the country semipermanently, taking advantage of modern long-distance telecommunications to edit and write remotely. Proctor's extensive travels and regular newspaper and journal pieces soon made him a national celebrity in the United States and likely a much more widely known and respected commenter on science than any other astronomer, including Newcomb. In January 1874, 1,800 people attended his sold-out New York lecture series, and the press coverage of his various pancontinental speaking tours was typically both extensive and laudatory.[103] Proctor was, in turn, a keen proponent of the United States' zeal for popular scientific knowledge and its relative lack of intellectual or class snobbery. After lecturing before 1,300 Bostonians across two sell-out lecture series, he described the Lowell Institute as "the first in standing, as it was the first in point of time, of all the institutions of the kind, not only in America, but in the world."[104] In private too he praised the "pleasant ways of scientific men in America," and upon his return to Britain he used the lessons he had learned on his travels as a stern warning for his staid compatriots. American "practical common-sense," he wrote, had advanced that nation's science and industry well past Britain in a wide variety of areas, from rail travel to meteorological forecasting to astrophysical observation. He could only rue the fact that the "true freedom of thought" he experienced in the United States was "tolerated only, not welcome in England, at present."[105]

"At present" was a crucial phrasing of the state of affairs. Like Stead, Proctor worked hard to Americanize Britain's attitudes toward egalitarianism and the democratization of knowledge. Lecturing the readers of the *Fortnightly Review*, Proctor used the example of John Lowell Jr.'s establishment of the Lowell Institute to pointedly contrast the profligate waste by "the ruins of monarchies . . . on war and spoliation, luxury and superstition" with philanthropic Americans' recognition of "the true use of science—the welfare and the culture of the many."[106] This was a particularly relevant example for Proctor, given the Lowell Institute's prominent place in a system of public education he knew well from firsthand experience: the American lyceum model of touring lectures. This system, like yellow journalism, traces its roots to the same emergent trends toward journalistic objectivity and individual self-improvement in the post-1830s American news economy. Public lectures were a vital component of this egalitarian knowledge order, with the American "cult of self-improvement" driving an explosion in public lecturing, especially on scientific subjects. Like the country's booming news economy, objectivity and inclusivity were both key facets of this enterprise. In contrast to the contemporaneous British exhibition and lecture scenes, American lectures were expected to embrace all members of a community regardless of social standing, and their content was intended to be both useful and nonpartisan.[107] "The characteristic of our age," lectured the liberal Unitarian preacher William Channing in 1841, "is not the improvement of science, rapid as this is, so much as its extension to all men."[108] Centered on community-funded lyceums, an extensive nationwide public lecture system developed that, as *Putnam's Monthly* noted, was unique to the country: "The lyceum is the American theatre. It is the one institution in which we take our nose out of the hands of our English prototypes—the English whom we are always ridiculing and always following—and go alone. The consequence is, that it is a great success. It has founded a new profession." Preeminent lecturers became both national celebrities and exemplars of the country's intellectual dispositions. "They are," *Putnam's Monthly* argued, "the intellectual leaders of an intelligent progress in the country. They are especially, and in the best sense, Americans."[109]

PROCTOR AND THE NEW JOURNALISM AT WORK

Proctor's ambition was to develop in the Old World what the diverse cohort of lyceum lecturers and science journalists had pioneered in the New. Reconfigured within this context, imaginative astronomy can be seen to emerge from both cutting-edge astrophysical research and the resources and constraints provided by the new journalism. Exemplary is Proctor's involvement as a founder member, in 1869, of London's Sunday Lecture Society (SLS). Modeled on the egalitarian lyceum model that Proctor had experienced on his huge lecture tours of the United States, the SLS aimed to promote in Britain the same democratic moral betterment that the lyceum system had pioneered in the United States. The society's stated mission was "to provide for the delivery on Sundays in the Metropolis, and to encourage the delivery elsewhere, of Lectures on Science,—physical, intellectual and moral,—History, Literature, and Art; especially in their bearing upon the improvement and social well-being of mankind."[110] As the liberal *Examiner* noted in its praise of the new society's work, in Britain there was still, "from the hard-worked artisan to the refined and cultivated lady of fashion," a profound and troubling ignorance of "the truths revealed by science." Science lecturing had declined in Britain after the 1830s, and Proctor and a cohort of like-minded lecturers were, the *Examiner* suggested, at the forefront of a long-wanted revival, making a "praiseworthy effort" to open the great book of nature and combat the "ignorance of the workshop, the counting house, and the drawing-room." This was a liberal—bordering on radical—project aimed at the masses and widely attacked for its anti-Sabbatarian ambitions. It therefore stood in stark contrast to London's longer tradition of high society gatherings at the Royal Institution and the spectacular lectures and displays at the Royal Polytechnic Institution and the Crystal Palace. Meeting "a want greatly felt for some time past," the SLS "will not have done all its work," the *Examiner* concluded, until it had established itself on the American lyceum model and "induced the establishment of at least some thousands of similar organizations all over the country."[111]

When Proctor wasn't lecturing, he was writing, and here too American influences loomed large. He was, for example, a regular contributor to one of Stead's main rivals in the London new journalism scene, the *Daily News*. Founded in 1846 by Charles Dickens to be a "Morning Newspaper of Liberal Politics and thorough Independence," the paper was an early adopter of new journalistic practices. In the early 1850s it had campaigned for the repeal of the Stamp Act, and it was one of the first successful post-repeal mass-market populist dailies, securing a circulation of more than 150,000 by 1870. Abolitionist and pro-North in the U.S. Civil War, the paper was a pioneer of telegraphic dispatches for the gathering of exclusive foreign news, entering into a syndicated arrangement with one of the boldest American proponents of yellow journalism, the *New York Herald*. In 1876 its vivid, shocking descriptions of the "Bulgarian atrocities" sparked a national uproar. Supported by Stead at the *Northern Echo*, this episode had established a turning point in state-media power relations, eliminating the government's traditional control over international news and ultimately dictating foreign policy (a technique subsequently much emulated by Stead).[112] As he later would for American equivalents like Joseph Pulitzer's *New York World*, Proctor wrote regular pieces for the *Daily News*, many of which were then recycled into one of his numerous collected essay books. These works spanned subjects well beyond astronomy, including natural disasters, technological innovations, social statistics, and the superiority of American preparations for the eclipse of 1870.[113]

By 1881 Proctor had become famous and wealthy enough to found his own independent journal, *Knowledge*, pitched as a populist rival to the publications of his enemies in the scientific elite, in particular Norman Lockyer's *Nature*. As Bernard Lightman has demonstrated, *Knowledge* very much bore the personal imprint of Proctor's editorial style, combining large quantities of content written by himself with an inclusive architectonic intended to draw his wide audience into the scientific arena and provide an open forum for egalitarian knowledge exchange. The format and style of *Knowledge*, Lightman notes, "reflected Proctor's aversion for the professionalizing hierarchical vision of science contained in the pages of *Nature*."[114] Though this format clearly owed a debt to the *English Mechanic*, several significant and novel practices adopted from the new journalism are worth highlighting. One is the central place in *Knowledge* for gossip, which, as Joel Wiener has stressed, is one of the most important but overlooked features of the Americanization of the British media in the 1870s and 1880s. Human interest content, particularly seemingly private tidbits, was exploited by populist editors as a way of stoking and satisfying their readership's often restless social curiosity. Gossip columns began to appear in U.S. papers from the 1830s, and their transfer to the British press was a central facet of the new journalism. Proctor would have been exposed to this kind of risqué editorializing through his association with the *Daily News*, which ran a popular gossip column, and he incorporated this style of human interest chitchat into *Knowledge*'s regular scientific gossip column.[115]

Beyond shared journalistic philosophies and methodologies, Britain's organs of new journalism and Proctor's *Knowledge* also shared interests in specific subjects, particularly those that cut across the themes of liberal political campaigning espoused by the former and progressive scientific debate promoted by the latter. In its first issue, *Knowledge* inaugurated what would be a long-running debate within the journal's pages over the alleged scientific evidence for the inferiority of women. The discussion was opened by an anonymous (and therefore almost certainly Proctor-penned) article that highlighted the "manifest weakness of the reasoning" behind arguments showing women to be at a lower stage of development than men.[116] The back-and-forth discussion that ensued through the correspondence pages of the journal serves as an archetype of the sort of open discourse on science that Proctor hoped to foster. It also mirrored Stead's own frequent agitations in the *Pall Mall Gazette* over women's rights and foreshadowed the fierce debates over "the sex question" and female suffrage that would ensue later in the decade, stoked in large part by the agitations of organs of new journalism such as the *Fortnightly Review*. In all three cases, the marriage of a hot topic to opinion pieces and ongoing (serialized) open discussion typifies the methods of new journalism and demonstrates the close marriage of political timing and commerce essential to the medium. For all three publications, the issue of women's place in society served as an ideal campaign ground on which to conduct the kind of debates that excited interest and sold journals in the highly competitive market of 1880s journalism.[117]

These links were explicitly recognized. Stead's *Pall Mall Gazette*, in particular, saw in Proctor a comrade doing for science what they were doing for politics and social justice. Expressions of mutual admiration are therefore telling of the shared social context in which these practitioners developed and exploited their new journalistic practices. As a catchall phrase for a wide suite of techniques, the new journalism was, after all, a phenomenon that, as well as its novel ability to invite speculation and engender heated debate, also flourished in part through rabid self- and cross-promotion.[118] It is telling, therefore, that Stead was at the forefront of promoting Proctor's work, especially once he had joined the astronomer in London in 1880. Stead's *Northern Echo* had already praised Proctor's "fertile pen," and the *Pall Mall Gazette*, once

Stead was on its staff, soon echoed such commendations. Proctor was, the paper wrote, a "delightful writer," an author of "perspicuous lucidity" whose many books and magazine articles "keep steadily up to the same high level of philosophical thinking, scientific precision, and literary excellence." Indeed, in the "due combination of these three elements" Proctor stood alone, "not only among his brother astronomers, but among the whole increasing clan of scientific popularizers."[119]

Proctor was for science what Stead advocated for journalism as a whole: a strong, combative editor who combined the traits of expertise, trustworthiness, moral rigor, and an exciting prose style. His great achievement, as the *Daily News* frequently noted, was one of combined practice and mediation—an exploration of the limits of human knowledge that by necessity also involved the skillful translation of esoteric knowledge to the exoteric sphere. "Mr. Proctor is not only one of the clearest writers who have ever expounded the discoveries of science to the unscientific world, but is himself an original and laborious investigator."[120] This rare gift to work effectively across these realms made him *more* of an expert than his more cloistered peers. He was, the *Pall Mall Gazette* effused, "first of all a thinker, then an astronomer, and finally a writer." Central to his success, therefore, was the ability to grasp myriad abstract concepts while at the same time deploying in his writing a synthetic, descriptive, visual, nonmathematical style. Proctor, the *Pall Mall Gazette* explained, "is no mere mathematical calculator of orbits and transits"; while he "kept constantly in the front rank" of his discipline, he also understood and could explain this vast erudition "in terms of concrete reality—to picture the objects vividly before his mind's eye in their actual relations, instead of envisaging them under the guise of x, y, and z, worked out in sheet after sheet of algebraic calculation."[121] For both the *Pall Mall Gazette* and Proctor, visual representations were heuristically and epistemologically superior to abstract mathematical formulations, precisely because they were, like their own journalism, democratic, synthetic, comprehensible, and informative.[122] Stead decried the tendency of specialist scientific journals to "emphasise division and isolation in science," and his newspaper praised Proctor for making "the common property of generations of professional astronomers the property of the ordinary English reader."[123]

The *Pall Mall Gazette* perceived too that Proctor's often bellicose agitations against the scientific elite accorded with its own campaigns against all forms of inveterate power. After Proctor had railed against the "false . . . profundity" of Arthur Cayley's abstruse 1883 BAAS

Presidential Address, the *Gazette* enthusiastically backed Proctor, dubbing him "the fighting man of the astronomers."[124] In throwing their support behind Proctor, the paper made clear that this was part of a broader campaign in favor of his model of participatory scientific practice and progressive disciplinary reform:

> As the evolutionist among astronomers [Proctor] has met with no little opposition from those scientific Gradgrinds of the old school who will hear of nothing but bare facts, and hate inference or generalization as illegitimate flights of imagination: opposition of the same sort as that offered to Mr. Wallace's first rough sketch of the natural selection theory, when sundry "eminent entomologists" complained that instead of propounding the doctrine which has revolutionized biology, he ought strictly to have confined himself to matters of direct observation. But for real philosophical grasp, luminous insight, and the power of mentally reconstructing cosmical history, Mr. Proctor has probably no equal among living brethren of his own craft.[125]

This was an ideological battle that explicitly linked the future of imaginative astronomy with science's—and therefore society's—progress. Its battle lines placed the expectations, actions, and responses of diverse consumers of astronomical content at the heart of the fight for the future of astronomy. How practitioners responded, therefore, must be understood as a trajectory of action shaped by these marketplaces. Revealing, for example, is the enthusiastic support for Proctor offered by the *Pall Mall Gazette*'s hero of biology, Alfred Russel Wallace. Writing to Charles Darwin to recommend two new books on the chemistry of the sun by Proctor and Mattieu Williams, Wallace presented his case explicitly in terms of imaginative astronomy—of speculation and progress versus cloistered dogmatism and authoritarianism. Proctor's and Williams's books had been "ignored by the critics," Wallace lamented, because, having encountered works "not . . . encumbered with any mathematical shibboleth," they had "evidently been afraid that any thing so intelligible could not be sound." The manner in which "*everything* in physical astronomy is *explained*," he continued, "is almost as marvellous as the powers of Nat. Select." and "naturally excites a suspicion that the respective authors are pushing their theories '*a little too far.*' . . . It does not say much for our critics that, as far as I know, [the books'] great merits have not been properly recognised."[126]

Yet such books did find large and critically engaged audiences whose tastes, reading habits, and responses must be accounted for.

For the late Victorian readers who the *Pall Mall Gazette* and Proctor targeted, comprised largely of people whose reading proclivities were outside the academic and intellectual elites, matters of "cosmical history" were of much greater interest and significance than matters of classical astronomy. With little mathematical education and limited money to spend on scientific products but with a commitment to self-improvement, such audiences sought out material that was lively, informative, and egalitarian.[127] Although they responded in diverse ways to the science that they encountered, these audiences tended to reject the roles of passive consumers of knowledge that elitist, professionalizing scientific naturalists attempted to confer on them. Active engagement with a style of science that was overtly participatory went hand in hand, therefore, with the kinds of social and political agitations that the new journalism championed. The influential socialist and trade-unionist Tom Mann credited Richard Proctor's writings and lectures as the major influence on one side of his twin abiding passions: "social problems and astronomy."[128]

From Proctor's perspective as a practitioner, this egalitarianism meant developing and promoting an imaginative astronomical discourse that was simultaneously participatory, engaging, and educational. Proctor's ability to "mentally reconstruct cosmical history," as the *Pall Mall Gazette* so evocatively put it, was fundamentally reliant upon an imaginative and exciting prose style, an authorial technique that developed and then played into audience expectations of entertaining and informative stories. It is in this context that we best see the roots of Proctor's most enduring contribution to the study of Martian astronomy and the plurality of worlds, his "new theory of planetary evolution."[129] This theory combined speculative deductions based on the findings of the spectroscope with novel and much-discussed ideas about thermodynamics and deep-time evolution. Proctor's aim was to present to his audiences, as he put it, "evidence in favor of a new theory of life in other worlds, and of the past and future of the solar system."[130] This expansive cosmic-evolutionary theory gripped audiences by reconfiguring worlds elsewhere as vibrant, animated realms, which ran a course "through burning childhood, fiery youth, manhood, old age and decrepitude, to the final stage—that of death."[131] In so doing Proctor constructed a complete narrative arc for Earth's neighboring planets, endowing each with life histories: Mars *might* already be a dead or dying planet; Jupiter *might* act as a sun illuminating life on its orbiting moons. It is precisely this tension between evidence and speculation in the service of a cosmic-evolutionary narrative that Proctor was playing on in his 1882 letter to the *Times* concerning the possible existence and nature of canals on Mars—a form of imaginative astronomy that would have been very familiar to most of the paper's readers.

As Simon Newcomb's critique quoted above makes clear, imaginative astronomy was an inherently visual methodology, intended to either construct mental images through words or to use pictures and photographs to convey the drama and beauty of the heavens. Much recent scholarship on Victorian popular science has stressed the centrality, indeed perhaps the primacy, of the visual in successful public expositions of science, a point much supported by Proctor's work.[132] As Lynda Nead has shown, this was a moment in which the new astronomical techniques of photography and spectroscopy were married with advanced printing techniques to produce and disseminate amazing new images of the heavens that collapsed time and space. "The camera and the telescope created images of a new order of sublimity that liberated the visual imagination and enabled it to roam through the immensity of the universe."[133] Making such images required an immense amount of work, often in the service of carefully constructing and managing multiple levels of abstraction. These technologies of representation therefore greatly problematized the sufficiency of the visual, exposing tensions between the visible and the invisible, the observed and the imagined, the objective and the subjective. With the limits of vision thoroughly exposed, some, including Proctor, posited the imagination as a viable tool for penetrating beyond the limits of what could be directly seen.[134] Proctor's writing and lecturing both created and then played off this new, much-contested visual order, and his books deployed written descriptions and speculative images that toyed with the reader's point of view, shifting perspective from a terrestrial viewpoint to a place somewhere else in the solar system. When considering Mars, Proctor wrote, "we might see in imagination the waves of those distant seas beating upon the long shore-lines . . . the slow progress of the Martian day—the mists of morning gradually clearing . . . the winds raised by the midday heat, zephyrs murmuring along the distant hills . . . the gathering of clouds toward eventide . . . leaving the same constellations we see to shine with greater splendour through a rarer atmosphere." Readers "can imagine all this" by placing themselves as an observer on the planet's ruddy surface because, Proctor confidently stated, "we know from what the telescope has revealed."[135]

Such techniques transcended the capabilities of photography, even when employing that technology. In Proctor's 1873 monograph, *The Moon: Her Motions, Aspect, Scenery, and Physical Condition*, readers were presented with albumen prints of photographs of the moon taken by Lewis M. Rutherfurd, presented alongside hypothetical illustrations that invited them to see and feel what it was like to become "lunarians" by standing beside craters and looking back at Earth. A year later, James Nasmyth and James Carpenter's masterpiece of imaginative astronomy, *The Moon: Considered As a Planet, a World, and a Satellite*, took this elision of photographic fidelity and imaginative transposition a stage further, illustrating the book with photographs of plaster models of the lunar landscape, shot as if the photographer were positioned on an adjacent crater or peak. Photographs of a wrinkled apple and an old hand were used to evoke the "origin of certain mountain ranges resulting from shrinking of the interior," and a section of the book took readers through "a day on the moon imagined." "Where the material eye is baffled," they wrote, "the clairvoyance of reason and analogy come to its aid."[136] Readers of Proctor's work and of Nasmyth and Carpenter's therefore shared a familiarity and comfort with this form of imaginative perception, built up analogically through a string of associations that could plausibly lead the reader from a photograph of a wrinkled hand to an image of the moon's mountainous topography (figure 1.2).[137]

This representation of astronomical phenomena as commonplace worlds was most spectacularly evident in the material culture of astronomical display. When Proctor first saw stereographic photographs of John Browning's Mars globe, he "could scarcely believe that they were not real Photographs of the Planet." In his booklet accompanying the stereograms, Proctor described the globe as "the exact reproduction in a solid form—or, I may say, the embodiment" of the observations by William Dawes. Proctor himself had used Dawes's drawings as the basis for his own map of Mars, and this new map would itself become the basis for the first mass-produced globe of the planet in 1873 (see figure I.1). Proctor cheekily promoted this new representation through one of his own (anonymously authored) articles, inviting his readers to imagine the appearance of Mars's "ruddy earth" and "greenish ice-masses of the so-called oceans" by "conceiving one of those Martian globes which Captain Busk has recently caused Messrs. Malby to make from Mr. Proctor's charts."[138] Such techniques of imaginative representation through globes, photographs, stereograms, hypothetical illustrations,

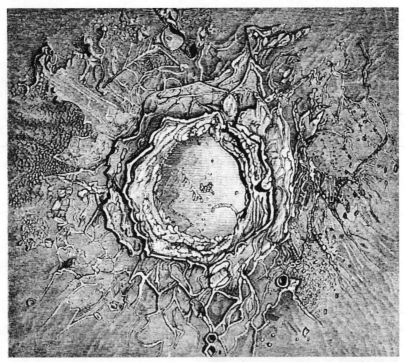

Figure 1.2(A–D). Four ways of representing the moon in the 1870s. A: the
"lunar crater 'Copernicus,' after Sec-
chi," as illustrated in Simon Newcomb's
Popular Astronomy (315). Conservative
authors like Newcomb continued to rely
on astronomical representations that
gave a grounded visual perspective, like
this hand-drawn illustration of a lunar
crater as viewed through a telescope.
B: a lunar photograph by Lewis M.
Rutherfurd entitled "Third Quarter,
Sept. 16, 1870, 1h. 49m. 0s. Sid. Time,"
as reproduced in Proctor's *The Moon*
(facing 230). Proctor called Rutherfurd
"the greatest lunar photographer of

the age," and he offered subscribers the opportunity to buy this book with larger
13 1/2 × 10 1/2-inch albumen prints mounted on card. C: a "lunar landscape, with
'full' Earth, &c.," also from Proctor's *The Moon* (facing 304). "In looking at one
of these views," Proctor wrote, "the observer must suppose himself stationed at

C

D

the summit of some very lofty peak" (304). D: a "group of mountains, ideal lunar
landscape," from Nasmyth and Carpenter's *The Moon: Considered as a Planet,
a World, and a Satellite* (plate 23). This work deployed a diverse range of visual
tropes to familiarize the moon, including using photographs of plaster models to
depict lunar topography from the perspective of a lunar observer.

and the new visual technologies of the illustrated stage show, appealed because astronomy, presented in this way, was both didactic and rousingly spectacular.[139] Nowhere is this clearer than in accounts of Proctor's great lecture tours, which regularly drew audiences in the thousands and earned him the status and wealth of a transatlantic celebrity. His lectures were "profusely illustrated" using the latest magic lantern technology, and his appearances on stage became cultural events in which cutting-edge data and spectacular words and images combined with a populist zeal for public education to produce the era's most successful scientific commodity: imaginative astronomy.[140]

Such performances must be understood within a broader print and "platform culture," in which the deployment of evocative written and verbal descriptions was just as important as the use of images for these representations' success.[141] The marketplace dialectic central to this imaginative astronomy therefore fundamentally shaped both Proctor's personal identity formation and the identity and success of his model of egalitarian practice. On the one hand, practitioners of this kind of astronomy fashioned for themselves a unique position as experts, claiming to represent both their science and the interests of their public, a process that required the careful negotiation of educative and spectacular imperatives through the skilled manipulation of visual and written representations.[142] On the other hand, readers, lecture-goers, and exhibition visitors, as participants in the production of a marketplace for this kind of astronomy, actively shaped the discipline's practice, never more clearly than in the case of Proctor's extraordinary success. The *Pall Mall Gazette*'s support for such an approach does not mean that the tastes of its audience were necessarily "lower" but rather that public taste for science mirrored the general trends that drew these readers to new journalism. Books and journals that were unpretentious, egalitarian, informative, socially conscious and liberal filled their bookshelves. An 1886 *Pall Mall Gazette* feature that tabulated readers' recommendations for holiday books included a progressive mix of Darwin, Dickens, Brontë, reformist social science (Wallace's *Land Nationalisation*, Henry George's *Social Problems*), and recent works of American comedy and drama (Mark Twain, Artemus Ward). It included only two blanket endorsements: the collected works of Charles Darwin and "any of R. A. Proctor's books."[143] Notably absent from the list were any titles from the X Club–controlled International Scientific Series (1871–1911), which, like *Nature*, attempted to marshal a top-down "popularization" of science. Placed within this context we can see that Proctor's model for his discipline, predicated on the construction of a planet or a star's "veritable life history" steeped in wonder and mystery, was a significant, compelling, and highly successful rival to that model of science being propounded by the X Club scientific naturalists.[144] Editors such as Stead and papers such as the *Daily News* were proving particularly successful at conducting print culture attuned to rapidly expanding late Victorian reading audiences, and Proctor's scientific program not only grew out of and mirrored their methodologies but also fed on the populist and democratic intellectual culture it fostered. As one *Pall Mall Gazette* review noted, "Science is nothing if not popular with Mr. Proctor."[145] To both Proctor and his readers, this was nothing if not a compliment.

This chapter has presented a new claim as to what constituted astronomical work between the 1860s and the 1880s (in the English-speaking world at least). Such work, I have argued, cannot be separated from practices of scientific communication. Problems and opportunities relating to the mediation of knowledge claims were embedded within and therefore constitutive of debates about how the discipline should conduct its work. My account therefore presents a rethinking of what change in astronomy in this era means. Neither popularization nor professionalization defined these transformations. Rather, it was the resources of journalism that constituted the era's central problematic. With a range of astronomies apparently viable, Proctor's imaginative astronomy successfully posited that science journalism represented the most plausible and fruitful way for the discipline to progress beyond its profound identity crisis. Read in this light, Proctor's 1882 letter to the *Times* is not, as has often been assumed, a striking and unprecedented intervention into a nascent debate about canals on Mars. Rather, it is an entirely typical intervention by the era's most widely read authority on the matter at hand, employing a style and tone that would have been generally recognized and read as characteristic and unremarkable by an audience used to (and receptive of) imaginative astronomy.

That Proctor's letter to the *Times* has subsequently been misread is simple to explain. The model for astronomy described here has been easy to miss because imaginative astronomy ultimately failed to establish itself within the toolkit of Proctor's discipline. Given the exceptional success and power of Proctor's program by the early 1880s, this failure needs to be understood too. Imaginative astronomy faced many challenges, and in the next chapter I consider these trials in terms of significant technological and spatial changes to the working environments and media relations of astrophysics. Just as the United States had been Proctor's exemplar, so too would this country be the progenitor of imaginative astronomy's downfall. It was here first that the contested relationship between astronomers, observatories, and mediums of mass communication would be decisively transformed. Massive technological and geographical developments that ran antithetical to Proctor's model for a participatory, egalitarian astronomy would inaugurate an era of select, high-cost, and remote astrophysical observatory science.

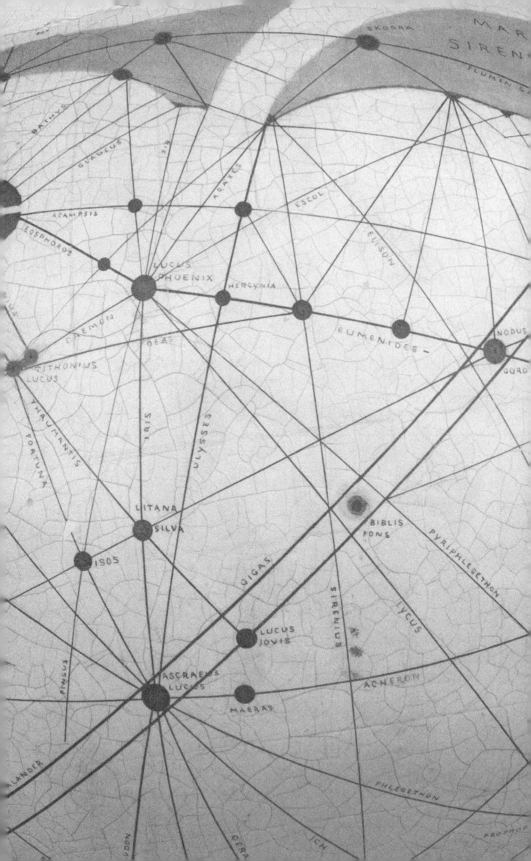

ANNIHILATING TIME AND SPACE 2

Observatories and the Technological West

In June 1886 Richard Proctor, the most famous astronomer in the transatlantic world, arrived in New York to begin another sell-out lecture tour. When he stepped off the Liverpool steamer, journalists tipped off to his arrival were already waiting dockside. The *New York Tribune*'s lucky reporter got the quote they were all hoping for: "I have come over to America to live you know," Proctor explained, "if I can stand the dry climate."[1] In truth, as excited as the *Tribune* was to report Proctor's immigration, it was only the last step of a process of American residency that had begun several years earlier. Although the reasons for this gradual move to the United States are not entirely clear, it is apparent at the very least that Proctor had a strong attachment to a country whose democratic intellectual tradition had so profoundly influenced his own work. More prosaic factors—personal tragedy, romantic entanglement, financial opportunity—also seem to have played a role in instigating his emigration from Britain. After marrying an American widower in 1881, Proctor eventually established homes in Saint Joseph, Missouri, and Manhattan, finally moving to Ocala, Florida, shortly before his untimely death from yellow fever in September 1888.[2]

Proctor's affinity for the United States was also likely exacerbated by his growing antagonism with colleagues in Britain. On April 14, 1882, the day after Proctor's letter on the Martian canals appeared in the London *Times*, Mars observer Nathaniel Green rose to speak on the matter at the RAS. While Green expressed respect for Schiaparelli's striking observations, he pointedly noted Proctor's suggestion of "very large engineering works" to mock the imaginative astronomy of his London colleague. "I do not wish myself to introduce any thing like a joke upon such a very serious matter," he sarcastically noted, "but I am persuaded that those appearances require great caution, that is, we should not recognize them as facts until others have seen repetitions of the same phenomena."[3] Green's own map of Mars, based on extensive observations by a network of British astronomers during the 1877 opposition, showed no canals and depicted the planet in a detailed, naturalistic rendering that eschewed hard contrast in favor of subtle shading. As K. Maria D. Lane has shown, British astronomers rallied in support of their compatriot's map, which Green himself advocated as an aggregated totalization of British astronomers' knowledge of the planet.[4]

Green and his compatriot's harsh criticisms also spoke of the rapidly entrenching positions of British astronomy's three-way disciplinary fight. While Proctor maintained a keen interest in extraterrestrial life, members of the Greenwich-Cambridge axis continued to do their best to exclude such questions from the purview of what Airy pompously called "the highest science."[5] Lockyer, meanwhile, read a more overtly elitist lesson in the furor over the canals, contrasting the absurd "imagination of the ready writer" with the salient lessons taught by "modern observers" such as Green. Pitching as always for the priority of institutionally certified astronomers, he caustically concluded that "the observations [of Mars] which engendered invention in one class of minds engendered doubts in others."[6] Proctor was becoming increasingly isolated among his British peers. Although there remained a large British audience for works that addressed Martian life, other than Proctor, with his continued transatlantic output, no other serious British astronomers actively engaged with the question in public. Robert Ball, royal astronomer of Ireland and an emergent rival to Proctor in the marketplace of popular astronomy, summed up this competing unimaginative philosophy when he declared that "the fact is . . . when an astronomer goes into his observatory for his night's work he finds it usually convenient to leave all the ecstatic and most of the poetic portions of his constitution outside."[7]

British astrophysical research took this mantra to heart. Lockyer's fierce campaign for a new state-funded rival to Greenwich had finally succeeded in 1879, with a dedicated solar physics observatory established in South Kensington and the *Nature* editor installed at its head. This limited Britain's interests almost exclusively to solar scrutiny, a focus that, as Proctor warned Airy, was "associated with the preposterous idea (worthy of Zadkiel and the editor of Moore's Almanac) that the phenomena of weather may one day be predicted from the records of solar spots, faculae, and prominences!" The sun became, according to Simon Schaffer, "a cult for British astronomers," with the tendentious links between solar astrophysics and what Lockyer dubbed the "Meteorology of the Future" the dominant concern.[8] Meanwhile in Greenwich, Airy's intransigence ensured that the Royal Observatory would only secure its own astrophysical observatory in 1899, under the direction of Airy's successor, William Christie. Elsewhere, Britain's few functional university observatories all variously failed to make any great headway with the spectroscope until the early twentieth century.[9]

This lack of growth of extra-solar astrophysics in Britain coincided with a remarkable flourishing of that subject in the United States. As historians of astronomy have stressed, the last two decades of the nineteenth century saw the United States first catch up with, then in many respects surpass Europe as the epicenter of astronomical science. This spectacular ascent to world-leading status was driven by the early, committed, and manifold adoption and development of astrophysical techniques. Accounts of this work have justifiably focused on the rapid advances made, in particular, in the realm of stellar evolution, where successive generations of American astrophysicists made groundbreaking discoveries and established world-leading reputations. What these histories tend to miss is that this ascent occurred at the same time that the United States also became the undisputed center of research into the question of life on Mars. Proctor's literal move to America was accompanied by a figurative one, in which not only discussions about Mars but also the next phase of research into the planet came to be dominated by American observatories.[10]

The geographical aspects of this American work provide one key explanation for this transition. As astronomical investigations shifted focus from the mechanical movement of celestial bodies toward these bodies' physical and chemical constitution, the nature of the observatory changed markedly. Observatories came to resemble laboratories, and with the exception of solar work they also came to depend upon expensive, sophisticated, and above all highly sensitive instrumentation, including massively powerful telescopes capable of gathering light from distant stars, nebulae, and planets. Geopolitics proved crucial in this new astronomical order. At the same moment that most British and European observatories were beginning to suffer from their metropolitan location and the spiraling cost of new telescopes, American philanthropists and entrepreneurs were endowing a new generation of super-expensive observatories, defined by both their remote mountain locations and their world-leading instrumentation.[11] These new remote sites were optimized for efficient data gathering and therefore carefully situated in locations with clear skies and calm air. Although their main nighttime duties focused on the observation of distant stars and nebulae, the periodic and brief oppositions of Mars offered important opportunities for these sites to exploit the exceptional penetrating power of their outsized telescopes. As recent studies have stressed, the exceptional quality of "seeing" afforded by the thin, stable atmosphere and minimal light pollution at these isolated observatories became a significant factor in debates over what astronomers thought they were observing on Mars.[12] Meaningful contributions to the debate about the canals therefore narrowed to a few prime observatory sites, predominantly in the western mountain ranges of North and South America.

This ascent of American astronomy to world-leading status, particularly in the spheres of astrophysics and planetary astronomy, is by no means self-evident, however. It is neither merely the product of shifting financial clout within the discipline nor a simple matter of moving astronomy wholesale to pure mountain peaks. As in the previous chapter, America's growing geographical domination of astronomy needs to be understood within the broader transatlantic context of changing disciplinary practices and their intimate relationship with changing communications technologies. To do this we must understand the place of American observatories, and to understand these places requires that we grasp the sociology of roles at play between astronomers, their disciplinary peers, and the mass media that they came to rely upon. As observatories moved west, I argue, astronomy's technological relationships with time and space were transformed, inaugurating both a crucial shift in the ways of seeing distant places and broader social interest in the power of long-distance, instantaneous communication. Mars would be the ultimate object for both these new paradigms, not just in the American context where they emerged, but more broadly in the transatlantic world and beyond.

PUBLIC ASTRONOMY AND PROFESSIONAL
CERTIFICATION IN THE UNITED STATES

The American context for these transformations differed in important ways from the British scene that Proctor had emerged from. Public and professional values in American astronomy had, since at least the middle of the century, been in a state of uneasy instability, as exemplified by a notorious controversy at Dudley Observatory. In early 1852 the United States' preeminent exponent of public astronomy, Cincinnati Observatory's Ormsby MacKnight Mitchel, spoke in Albany, New York, to encourage the city to construct an observatory alongside the university they were then planning. Mitchel had been an early influence on Proctor, and his model of a publicly subscribed and open Cincinnati Observatory constituted a novel republican prototype for scientific practice.[13] But when its first copy, the Dudley Observatory, opened in Albany in 1855, Mitchel turned down an offer to become its director, leaving its board of trustees scrambling to appoint an alternative. This led them into the arms of the Coast Survey's Alexander Dallas Bache and his tight-knit cabal of elitist friends, the self-dubbed "scientific Lazzaroni."[14] Bache duly maneuvered the German-trained astronomer Benjamin Apthorp Gould into the directorship, and Gould in turn soon appalled the observatory's trustees with his prickly personality and his disdain for anything approaching public engagement. In the pamphlet war that followed, Gould paraphrased his hero, the doyen of German positional astronomy Friedrich Bessel, declaring that "the science of astronomy consists in the investigation of those laws which govern the motions of the heavenly bodies. The Natural History of the Heavens, although a cognate and most interesting field of inquiry, is not included in the domain of astronomy proper." Under attack because he would not let the public into his publicly funded observatory, Gould reminded the trustees that "no principle is better established than this . . . that the abstruseness and want of popular interest of investigations increase in proportion to their scientific value."[15]

Historians have pointed to the Dudley controversy as exemplary of endemic "hostilities and jealousies" among the American astronomical community and have proffered such divisions between practical and physical astronomers as a reason for the discipline's failure to establish a nationwide professional society until 1899.[16] The intricacies of this disaccord therefore provide considerable insight into the nature of dramatic changes in American astronomy in the last half of the nineteenth century. As with all American science, this was a period of dramatic growth, heralding changes that have often been considered under the problematic rubric of professionalization but which have more recently been analyzed through the more nuanced and dynamic categories of expertise, power, and localized social forces. As Nathan Reingold has noted, it was not a simple binary conflict over amateur or professional status that characterized such changes so much as a multi-faceted and contingent contest for "certification . . . the seeking and attainment of place, income, and influence."[17] These complex struggles for certification are revealing precisely because the mechanisms for its attainment were far from clear. Particular to midcentury American astronomy (at least in contrast to Europe) was a notable *lack* of organizational and disciplinary structure. This lack left a considerable amount of room in which to develop working models of what "certification" might constitute.

Unlike in Britain, there was no relatively centralized controlling body or established metropolitan elite. Even when such institutional hierarchies began to form, ongoing tensions emerged from attempts to reconcile specialization with the egalitarian proclivities of American society.[18] The professionalizing Lazzaroni might look at first blush like an Americanized version of the X Club scientific naturalists, but, as Reingold notes, the men behind institutions like the American Association for the Advancement of Science and the National Academy of Sciences were "raving egalitarians" in comparison to their British equivalents. "Even if they had not been infected with republican or democratic principles," he concludes, "a continuing establishment control was not feasible given the size and complexity of the USA." This was particularly true for the field of astronomy, which was unusually large—144 observatories were in operation by 1882—and extremely diffuse. Pioneering work relating to land mensuration and westward expansion had developed a "geophysical tradition" in which astronomy, meteorology, navigation, and geophysics were intermingled within large-scale government projects, creating a "research ideal" with a "blurred boundary between theory and practice." And when this geophysical tradition began to wane after 1860, the rise of novel astrophysical techniques imposed new questions about the appropriate place of experimentation, public engagement, and precision measurement within the discipline. These questions were amplified by a broader post–Civil War trend away from the attitudes of individualism and generalism that had typified intellectual life before 1860 toward an appreciation of professionalism, expertise, and scientific method.[19] Conflicts such as those at the Dudley Observatory can, therefore, only be understood as symptoms of a particularly American struggle over the nature of both astronomical work and social status in a rapidly growing and almost unbounded discipline.

From the perspective of a transatlantic star like Proctor, the doctrines of imaginative astronomy must have appeared a particularly good fit for restructuring the United States' discordant and disorganized astronomical community. Certainly, from relatively early in his career, the Londoner worked hard to promote in the United States just the egalitarian and participatory scientific ideals that he himself had adopted and adapted from numerous New World influences. As early as January 1874, Proctor dined with eminent guests at a reception in his honor in New York, held to celebrate the launch of another sold-out lecture series. More than 1,800 tickets had been snapped up in New York alone, and demand had so outstripped capacity that the celebrated orator had agreed to deliver extra lectures in the city at the conclusion of his three-month pancontinental tour.[20] Yet this mass appeal also attracted the ire of American opponents of a less imaginative bent. It would be this cohort of European-minded (and often European-trained) astronomers who would carry the fight against Proctor on American soil.

At their vanguard was the redoubtable figure of Simon Newcomb, the navy astronomer we first met in chapter 1 viciously savaging Proctor's 1874 "great run" of lectures. Born in Canada and largely self-educated, Newcomb was a prodigious mathematical talent with an insatiable work ethic and a staunch commitment to the most abstruse aspects of positional astronomy. As the head of the U.S. Nautical Almanac Office he had systematically overhauled the theoretical and computational bases of the American Ephemeris, producing new astronomical constants that were ultimately chosen as the international standard for almanac work. Reminiscing on this achievement in later life, Newcomb modestly declared that those men who had, as he had, tackled the problem of the fundamental astronomical constants were "in intellect the select few of the human race,—an aristocracy ranking above all others in the scale of being. The astronomical ephemeris is the last practical outcome of their productive genius."[21] Like George Biddell Airy in Britain, Newcomb was in the unusual position of being a public positional astronomer, and among his New World peers he was considered North America's preeminent scientist. Even more than Airy, he leveraged the authority his post and his reputation had secured for him to exert considerable public influence, fashioning for himself a civic persona built in part upon a resounding repudiation of any imaginative tendencies in the astronomical worker. Rejecting the very premise of Proctor's disciplinary model, Newcomb sternly informed his American public that, "so far as astronomy is concerned, we do appear to be fast reaching the limits of our knowledge. . . . The work which really occupies the attention of the astronomer is less the discovery of new things than the elaboration of those already known, and the entire systemization of our knowledge." For Newcomb, astronomy approached completion. Of the few loose ends that remained, mathematics, not imagination, held the key to their solution.[22]

Despite their considerable differences, Newcomb's imprimatur still mattered to Proctor if the British astronomer were to establish a professional reputation stateside. In private, the two men's surviving correspondence suggests at least an initial attempt at cordial professional relations, though this very quickly broke down. The flashpoint for the falling out was an anonymous review of several of Proctor's books published in September 1874 in the high-minded American literary periodical the *Atlantic Monthly*. This review, and the fallout from it, reveals a powerful American alliance against Proctor that would prove influential in the nascent professional organization of U.S. astrophysics. The *Atlantic*'s review was certainly harsh, accusing Proctor of issuing too many books that were all too popular, "so that our ears ring with stories and guesses about meteors, comets, colored suns, Mars and Jupiter . . . about these same depths 'astir with life.'" Just as he had in Britain, Proctor found his status as an astronomer questioned because of his authorial proclivities. Why, the reviewer asked, "should it not be possible for Mr. Proctor's great talents to be usefully employed on *real labor*?" Carelessly, however, the reviewer had singled out for ridicule a book that, though announced, Proctor had yet to actually publish. Seizing on this rather large error, Proctor immediately sent a fierce letter of response to the *New York Times*, announcing to its readers "an offense . . . committed against the morality of literature." Proctor's fame quickly propelled the affair into the transatlantic press, and the *Atlantic*'s anonymous reviewer was soon being dragged through the mud by a wide range of papers and journals.[23]

Meanwhile, Proctor took the matter up with Newcomb in private. Although the Washington astronomer was an obvious candidate to have written the review, Proctor appears from an early stage to have suspected one of Newcomb's close colleagues as the offending author. Either way, it was a sensitive matter, given Newcomb's power. Aware, no doubt, that his austere colleague was unlikely to be predisposed to his case, Proctor fished for sympathy by angling to portray a characteristically American style to his work: "I know you think some ideas I have thrown out incorrect. . . . But though our ways of viewing these matters and indeed the whole subject of astronomy, are unlike (for I, in turn, have my prepossessions—very strong, too—against routine astronomy, *at least as developed in Europe*) I had hoped that you would perceive that I take the known laws of physics as affording the only safe rule for judging about the unknown."[24] European habits were a malady, not a model. Just as he had in his manifesto, *Wages and Wants of Science-Workers*, Proctor pointedly used the epithet of "routine astronomy" as a means of inverting the disciplinary hierarchy, exposing in the *Atlantic*'s tone a serious failure of not just the morality of literature but of American astronomical labor as well. "Of course it is a 'trouble' to put scientific results popularly," he lamented to Newcomb. "Still, as none but the worker can really do so properly, it is well for those who can to make a little sacrifice in that direction—as Tyndall, Huxley, Airy, Darwin, and others on the other side have done, but very few (so far as I know) on this [U.S.] side."[25] Lured in by the abstruse mathematics of august European observatories, American astronomers ran the risk of learning precisely the wrong lessons from their Old World counterparts. Indeed, it was exactly this bad education that had enabled Proctor to identify his wayward critic. Writing to Newcomb with a tone of barely concealed satisfaction, he laid out the evidence: "I knew (i) that the critic must know you, for he quoted a remark made to yourself by Winnecke, (ii) that he was young from the general cheekiness, (iii) that he was rather Germanised, (iv) that he had probably been at one time under quasi military discipline; *und so weiter*."[26]

One man fit the bill: Edward Holden. West Point trained, disciple of the mathematical astronomer William Chauvenet, and recent assistant to none other than America's doyen of Germanized astronomy, Simon Newcomb himself, Holden was the immaculate antithesis of everything that Proctor stood for. His exposure was, to say the least, an awkward one for Newcomb. Holden had been his understudy and was therefore directly attached to his work, forcing the Washington astronomer onto the defensive. Nonetheless, he held fast, defending Holden from Proctor's fierce condemnation despite fears that Proctor might go public with their correspondence.[27] Newcomb's resolute defense of his impertinent assistant is therefore quite revealing. Holden was certainly no mathematical talent, and clumsy errors like his "discovery" of a third satellite of Mars exposed him and his superiors to a degree of professional ridicule, not least from Proctor himself.[28] Newcomb's patronage was, as such, much more a matter of disciplinary stewardship than professional nurturing. Indeed, it was the growing success of imaginative astronomy itself that forced Newcomb to take an active interest in the careers of men like Holden, who were better fitted to the management of new astrophysical observatories beyond the immediate ken of navy control. Importantly, Holden had proven himself to be both loyal and an aggressive defender of skeptical and unimaginative labor. This made him at least an ideal acolyte to fill one of the suddenly multiplying positions then emerging to service America's observatory boom. Of principal concern to Newcomb was not astronomical proficiency per se so much as the need for sound future management of these new institutions, which meant, above all, copying regimes of labor and discourse from the rational and exacting model of European positional astronomy. Hence the culmination of Newcomb's long patronage of Holden was the latter's appointment in 1885 to the directorship of the as-yet-unbuilt Lick Observatory in California, on Newcomb's direct recommendation.[29]

Newcomb was a navy man, and the Lick was a private enterprise. His key role in administering the most expensive and lavishly equipped astrophysical research institution ever built only makes sense, therefore, on these terms—as a direct response against less austere models for the future of American astronomy. Newcomb implicated himself into the emerging union of physics and astronomy precisely because he had an important stake in its correct stewardship. On the one hand, his patronage of work such as Albert Michelson's experiments to determine the speed of light promised the generation of new precision standards directly relevant to his beloved ephemerides. But beyond such immediate gains, there existed a wider interest among a close-knit group of astronomer-managers to secure for their nation's new astrophysical institutions exactly the same moral and practical norms that had brought positional astronomy to near completion. Young, ambitious, European-facing men like Michelson, Henry Rowland, Edward Pickering, and Edward Holden advocated for the endowment of observatories that were at once lavishly untraditional in their laboratory composition and yet took their labor model very much from the cloistered, elite, rigorously controlled institutions of European positional work. "That simplest of the departments of physics, namely, astronomy, has now reached such perfection," Rowland explained to the American Association for the Advancement of Science, "that nobody can expect to do much more in it without a perfectly equipped observatory; and even this would be useless without an income sufficient to employ a corps of assistants to make the observations and computations."[30] Much to Proctor's growing disdain, just when he was attempting to establish an Americanized epistemology of egalitarian science in the transatlantic world, a new generation was trying to do exactly the opposite, that is, to promote the incorporation into American astronomy of an ultra-high-cost, elitist, "Germanized" ideology of cloistered research. Simon Newcomb was their godfather.

OBSERVATORIES, NEWSPAPERS, AND THE TELEGRAPH

Proctor's move to the United States therefore coincided with a pivotal moment in the growth of American astronomy. The divide between a public astronomy centered on the natural history of the heavens and a professional astronomy centered on precision measurement appeared to be reconfiguring itself through an alliance of Germanized ideals and cutting-edge astrophysical technique. At stake was the place of the observatory in American public life. Just as with the British case discussed in the previous chapter, changes in the scope and impact of communications technologies would prove crucial. Most directly, the diversity of platforms available for communicating work fostered uncertainty over disciplinary identity and engendered heated conflicts over boundary formation. Attempting to impose order on their nascent subdiscipline, astrophysicists made a complex set of discursive moves intended to establish a particular kind of personal identity, protect autonomy, correct procedure, moderate disputes, or sanction only some kinds of knowledge as real science.[31] After his enemies successfully hounded him out of the RAS, Proctor moved to the United States no doubt in part because he was enamored with the greater diversity and liberality of the American scientific scene. Yet here too he found elitist critics willing to demarcate his work as outside the appropriate boundaries of astronomical labor.

Unlike in Britain, the geography of that work mattered a great deal too. In the vast territorial expanse of the continental United States, contests over appropriate rhetorical technologies and genre choices were inextricably bound up with the ways in which such knowledge traveled. Without effective centralized control and with such a large and diffuse geographical distribution, each American observatory was an independent agent in a marketplace of ideas. The United States had nothing even approaching the RAS or the Greenwich-Cambridge axis, so to attain certification American astronomers had to project authority by establishing the value of their local endeavors. This projection of authority required the translation of observations and astronomical news from a local to a national and international setting, a far from simple task in such a diverse and unbounded disciplinary landscape. The contrast between the approaches of Cincinnati Observatory's Ormsby Mitchel and Dudley Observatory's Benjamin Gould is telling, as it indicates a long-running discord over how one might go about doing this. Mitchel was a master at raising his profile locally to secure money for the observatory he worked to make internationally famous as a center of popular astronomical teaching. Gould, in contrast, demanded that he be left alone to raise the profile of the Dudley Observatory through the long-winded, systematic accumulation of positional data that could then be fed into the European ideal of "timeless" watchtower observatories. Both men had to answer to local trustees and subscribers, who themselves often had divergent ideas as to why they were funding astronomical work. Mitchel lamented that "the sublime wonders of the heavens . . . were powerless in their influence upon the minds of the multitude" for as long as they had been "confined to the hands of the rich and powerful, or locked up within the walls of institutions whose thresholds were never profaned by the footsteps of the people." Gould, meanwhile, ridiculed the American public's tendency to remunerate the "mere bookwright" rather than the "scientific investigator."[32]

Because such contests for money, publicity, and certification required the projection of local work and values over considerable distances, developments in long-range communications technologies would prove hugely significant. For both Mitchel and Gould, getting the word out or defending one's reputation were tricky propositions given their observatories' locations away from major metropolitan centers. Mitchel was a master of utilizing books and lectures to raise his institution's profile, and he founded his own journal to promote his observatory's work.[33] Gould resorted to pamphlets and the behind-the-scenes machinations of his Lazzaroni allies when he needed to ridicule the trustees of his observatory. Yet from the 1870s onwards, these technologies, though they remained useful, became subordinate to newspapers as the principal mechanism for distributing astronomical news. Above all, it was the rapidly expanding international telegraph network that underpinned this shift. As several impressive studies of fin-de-siècle science and culture have shown, this powerful new technology was integral to a profound reorientation of time and space in the last decades of the nineteenth century. Particularly in the United States, where distances were so vast and astronomy was so unbounded, this reorientation had significant implications for how astronomers managed their struggles for authority and certification.[34]

Several developments placed American observatories at the heart of an international telecommunications network. As studies of geophysics and standard time have stressed, astronomers began to connect their observatories to local and then national telegraph systems from midcentury, both to sell time to local railways and businesses and to participate in the accurate determination of longitude via the "American method" of telegraph signaling.[35] With the establishment of priority in new discoveries (particularly of comets) long an important issue for astronomers, the telegraph also quickly became a vital tool for the distribution of news of special celestial events. From 1882 Harvard College Observatory in Cambridge, Massachusetts, took over from the Smithsonian Institution in Washington, D.C., as the American center for coordinating the circulation of this news to a global network of observatories.[36] But a direct telegraphic link was far from sufficient for the management and projection of astronomical news. Establishing the value of local endeavors and securing credit for the generation of new and interesting knowledge also required that such information be visible beyond the walls of the observatory. So astronomy was news in the literal sense, as a form of discourse wedded to forms of mass communication, above all daily (or twice-daily) newspapers. When Norman Lockyer asked Simon Newcomb for advice on securing publicity and funding for an eclipse expedition to the Unites States, the Washington astronomer responded with the advice that "you make some noise in the journals about your coming. As your plans are developed telegraph them to our press, a sentence at a time. This will help in every way."[37]

In this way cable networks rapidly became a technology central to the astronomical news economy. News agencies such as the Associated Press, which were a direct product of the telegraph system, generated stories that were carried across hundreds of newspapers, and local news that originated in one paper, if deemed significant enough, would soon be cut, transmitted, reproduced, and excerpted by numerous others, often within hours. News as a commodity replaced political comment and local mercantile bulletins as the principal currency of the daily paper.[38] With astronomy already a recognized part of the news economy, telegraphically connected newspapers soon established a symbiotic relationship with observatories distributed across the United States and beyond. Harvard College Observatory, for example, cemented these ties by handing over the administration of its telegraphic news distribution service to John Ritchie, a science journalist who wrote for various Boston newspapers. "Every daily paper," noted *Popular Astronomy*, "is anxious for the news concerning current astronomical events, eclipses, discoveries of comets, erection of observatories, etc., while many make the publication of astronomical articles a regular and prominent feature."[39]

Ironically, the advent of this close relationship between American observatories and newspapers threatened to push Proctor, long the master of newspaper astronomy, into a marginal position. Whereas in London his status as an astronomer unaffiliated with any observatory had been a positive virtue (at least in the eyes of antiestablishment allies like the *Pall Mall Gazette*), in the United States Proctor's independence jeopardized his access to the national news network. It is telling, therefore, that in December 1884 Proctor quietly tried to secure for himself the directorship of the Drapers' private observatory in Hastings-on-Hudson, New York. Two years earlier, Henry Draper, the observatory's founder, had died suddenly from double pleurisy at the age of forty-five. Draper had been a pioneer of astrophotography and a powerful adversary to Proctor's British astrophysical enemies, particularly Lockyer and Huggins, and his widow and longtime astronomical collaborator Anna P. Draper now sought a suitable replacement to carry on this work.[40] In proposing himself for the job, Proctor very much played up his credentials as an astronomer capable of communicating Draper's and the observatory's work to an international audience. "No one living I am sure," he wrote Anna, "has a more earnest wish to see justice done to Dr. Draper's work. . . . I think I may add that few would be better able, and none could be more anxious than I, to make his work known and appreciated." Anna Draper soon rejected these overtures, however, on the very reasonable grounds that Proctor lacked the necessary experience to manage an astrophysical laboratory. This left him institutionless as ever and therefore in need of forging strong relationships with newspapers local to his various residencies.[41] Cut off from the growing network of U.S. astrophysical observatories, Proctor instead reasserted himself as a vibrant rival to the more deleterious practices of these institutions. The result was an astronomical and a newspaper sensation.

Proctor's target was a long-standing enemy, the author of the vicious *Atlantic Monthly* review, Edward Holden. By late 1886 Holden was preparing to take charge of a working Lick Observatory following a protracted and expensive construction process. The observatory had been endowed in 1874 by the eccentric land baron James Lick, whose seven-hundred-thousand-dollar bequest had stipulated that it should possess a telescope "superior to and more powerful than any . . . yet made." Finally, on December 27, 1886, amid much press fanfare, the largest and most expensive telescope lens ever constructed was delivered to the observatory's remote site on Mount Hamilton in Santa Clara County, California.[42] Proctor was not impressed. Though a new and particularly American paradigm, this alliance of extravagant philanthropy, careerist astronomers, and a giant, high-cost telescope constituted only a new iteration of Proctor's long-standing disdain for big money and big institutions in his discipline. That the project was headed by Holden only further excited his ire, and he duly took aim through one of his several newspaper connections in New York, writing a withering assessment of California's "big lens" for Joseph Pulitzer's *World*. Like Proctor's closest British ally, the *Pall Mall Gazette*, Pulitzer's *World* was an innovative, democratic, populist paper that melded novel design with probing interviews, sensational reporting, and an egalitarian zeal for taking on inveterate power. A pioneer of the "yellow journalism" that Pulitzer's name is now synonymous with, the *World* mixed populist crusades—such as Nellie Bly's notorious exposé of Blackwell Island's infamous asylum—with heavy doses of self-promotion, including one publicity stunt in which Pulitzer proposed erecting a giant billboard visible to readers on Mars.[43] No doubt the *World*'s populist tone and massive circulation appealed to Proctor, and the chance to stir up an astronomical quarrel appealed to Pulitzer and his deputy editors.

Published under the provocative headline "PROCTOR ON THE BIG LENS: HE THINKS THE LICK TELESCOPE WILL DISAPPOINT SCIENCE," Proctor's article was framed around the premise that once the fanfare around their mounting had faded, most recent giant telescopes had tended to disappoint. As he knew better than most, the two decades preceding the Lick's construction had seen a remarkable run of struggle and often failure with giant instrumentation, with new massive telescopes in Malta, Newcastle, Edinburgh, Toulouse, Paris, and especially

Melbourne all failing to live up to high expectations. The Great Melbourne Telescope, the joint-second largest reflector ever constructed, had been dubbed "a gigantic philosophical blunder" by the Australian press, and its abject failure in contributing to debates over true nebulosity had been thrown into stark relief by William Huggins's triumphant work on this question using a spectroscope attached to a very modest eight-inch refractor.[44] Though Proctor conceded in his *World* article that "theoretically a large telescope is capable of doing better work than a small one," such estimations of their value disregarded "the man at the small end of it." As far as Proctor was concerned, ample evidence already existed to prove that expertise as an astronomer trumped access to big-money equipment. "When I am asked what science may hope or expect from the great telescope of the Lick Observatory," he concluded, "I feel constrained to reply that, judging from recent experience, we can hardly expect aught but disappointment."[45]

Implicit in such a criticism was the suggestion that the Lick lacked a competent astronomer to mitigate against the myriad challenges of giant instrumentation. This called into question both the Lick's spectacular new telescope and the abilities of its director-in-waiting, a state of affairs that did not go unchallenged. Tellingly, it was the Lick's local populist newspaper, the *San Francisco Daily Examiner*, that led the response. Only a week before Proctor's article had been published, editorial control of this paper had passed from the millionaire mining magnate and U.S. senator George Hearst to his Harvard dropout son William Randolph. The younger Hearst had discovered a passion for journalism after working under Pulitzer at the *World* in 1886, and he had determined to turn around the fortunes of the *Examiner* along exactly the new journalistic lines that Pulitzer had pioneered.[46] So it was that one of the very first journalistic sensations orchestrated by William Randolph Hearst was a fight with Richard Anthony Proctor. Collusion with Pulitzer seems to have played a part, Hearst's paper having received Proctor's piece a day before its publication in New York, enabling the *Examiner* to respond the very same day. That both papers seemed to have been more than happy to stoke the flames of a scientific dispute speaks to the complex and increasingly challenging nature of the now-necessary relationships forged between observatories and their local agents of news distribution. The most conspicuous effect of the sensation that ensued was the projection into the public sphere of Proctor and Holden's long-standing mutual enmity. After the *Examiner* opened the conflict by dubbing Proctor's attack on the Lick "a proposition of self-evident absurdity," the paper then ran a long interview with Holden in which he equated Proctor's inability to grasp "a known fact" about telescopes—that bigger meant better—with the fallacious claims of the crank weather prophets E. Stone Wiggins and Henry G. Vennor. This was boundary work of the most direct kind, and Holden made his point quite explicit: "Mr. Proctor . . . *is not a working astronomer, but a writer of books.* He has done much to popularize astronomy in the way of compilations and for this deserves credit. He has made the science familiar, but the telescope used by him was a small one, and he has never yet done anything of consequence."[47]

As was to be expected, Proctor soon fired back. His letter of reply to Holden was gleefully printed by Hearst under the banner headline "A SCORCHER," with its main revelation—that Holden had been the anonymous author of the "spiteful and dishonest" 1874 *Atlantic Monthly* review—garnering the Californian astronomer much-unwanted attention. Transposed to the newspapers and journals in Proctor's homelands of New York and London, the argument now turned toward the personal character and credit of the Lick's house observer.[48] Here, the moral geography of astronomical labor was front and center. As the *New York Tribune* put it: "All the Pacific Coast scientists are after [Proctor] hot-foot, and they are saying the most spiteful and uncomplimentary things about him in the California papers." W. W. Payne, editor of the *Sidereal Messenger*, an American general interest journal of astronomical news, entered a note on "The Holden-Proctor Unpleasantness" to back the latter unequivocally. "If true and fairly presented, Mr. Proctor is right, in just severity and dignified silence both of which he has chosen to use." Payne's journal was read widely by astronomers on both sides of the Atlantic, leaving Holden with no choice but to reply, sending the same letter that he had sent to his local newspaper after Proctor's initial attack.[49] Holden, by this stage thoroughly outmaneuvered by his much more media-savvy opponent, could only look on as his reputation was given a sound kicking. Proctor, having started by focusing on the utility of new, giant telescopes, deftly transformed what was now a transatlantic sensation into a lesson about both Holden's individual incompetence, and the entire moral economy of astronomical labor:

> I am not greatly concerned to point out that in writing books which
> will sell because simple (though sound), I am adopting one way of
> earning a livelihood, while a man who takes a position in an obser-

vatory or a professorship at college takes another, the only question affecting the propriety of any such course being whether the work is well done, and my livelihood absolutely depending on my doing my work well, whereas, as Prof. Holden's case at Washington shows, a man may be inefficient as an observer and yet retain a handsomely paid official post in an observatory.[50]

Giant telescopes, Proctor suggested, were only the most visible facet of a corrupt new order in astronomy, predicated on the dubious alliance of big-money philanthropy and high-salary careerist astronomers. Holden's attachment to the Lick made clear that this new order was undergirded by the patronage and influence of just those "German-ized" astronomers that Simon Newcomb had endorsed and Proctor had railed against. In this fight, Proctor's last and only hope was, as ever, the eastern metropolitan newspapers still sympathetic to his alterna-tive, populist program.

From Proctor's perspective, early 1887 was therefore a moment of crisis for his discipline. The delivery of the Lick's giant new lens into the hands of an incompetent acolyte of Simon Newcomb was only the most public episode in a troubling new trend. In the very same month that Proctor's Lick article had appeared in the *World*, Uriah Boyden's $230,000 endowment for the establishment of an astronomical obser-vatory at altitude was handed over to another thoroughly unimagina-tive and salaried astronomer, the director of Harvard College Obser-vatory Edward Charles Pickering. Only a few months later, Edward Spence, trustee of the University of Southern California, announced plans to subscribe from wealthy benefactors a new observatory sited on Mount Wilson near Los Angeles, California, and equipped with a telescope even larger than the Lick's.[51] Proctor's intervention therefore speaks of a key moment in the transformation of American astronomy. With the Lick almost complete and two further massively expensive mountaintop observatories in planning, the model for astronomy that he had based his career on looked to be in serious jeopardy. As much as this was a contest over the structure of astronomy as a discipline or the moral purpose of astronomy as an enterprise, it was also fundamen-tally a contest over the *place* of astronomy. Holden and his careerist colleagues' rise to power was intimately connected with the consoli-dation of astrophysics in a few remote, expensive astronomical sites. As technologies of communication enabled the projection of authority over vast distances almost instantaneously, astronomers able to secure suitable funding could begin to move away from metropolitan centers in search of better environmental conditions for their new, massive telescopes. This confluence of money, technology, and environment occurred in the American West.

THE RISE OF THE TECHNOLOGICAL WEST

The alliance between Holden, the San Francisco press, and the Lick Observatory reveals precisely the union of science, massive technology, business, and publicity that characterized what Rebecca Solnit has called the "technological Wild West." From midcentury, Solnit shows, pioneer scientists, explorers, and businessmen in the western United States pushed the limits of high-cost technoscientific endeavors and in so doing constructed a new, technological West "as a distinct culture."[52] This conquest and transformation of the American West was fueled both by industry's demand for natural resources and by the technoscientific advances—in communications, industry, and warfare—that made such rapid capitalist and geographical expansion possible (see figure 2.1). The massive wealth of "robber baron" entrepreneurs such as Leland Stanford grew out of projects like the transcontinental railroad, the complexity and scale of which dwarfed eastern manufactures. Much of the massive wealth generated by these endeavors was then plowed into projects aimed at bolstering the reputations of their benefactors and glorifying California and the Midwest as a unique and heroic place.[53] Cutting-edge scientific research and its institutions were central to this project. Stanford's philanthropy supported the revolutionary motion studies of the photographer Eadweard Muybridge, as well as the university that bears its benefactor's name. Likewise, seven hundred thousand dollars of James Lick's vast wealth, accumulated through gold rush land speculation, was ostentatiously endowed for the construction of his namesake observatory. Lick's scientific monument was then handed to the University of California as an emblem of its state's superiority. Through "this act of spontaneous, and almost unparalleled munificence," noted the university's regents, "the interests of science . . . may be able to attain upon the shores of the Pacific and in one of the youngest states of the Republic, a higher advancement than has ever yet been reached by the oldest and most enlightened nations of the globe."[54]

Two abiding characteristics of these projects stand out: a transformation in humankind's relationship to the natural world and, with it, a radical alteration of human perception itself. Leland Stanford's

Figure 2.1. "American Progress," by John Gast, 1872. In Gast's now-famous allegorical representation of the conquest and modernization of the American West, the spirit of progress, Columbia, strings up a telegraph wire as she moves west, accompanied by soldiers, miners, pioneer farmers, and an advancing railroad, as well as fleeing American Indians at the edge of the frame. Reproduced courtesy of the Autry Museum of the American West.

railways and Eadweard Muybridge's "instantaneous" high-speed photographs, along with the telegraph wires that broadcast their achievements, shared the ability to transcend human limits of speed and perception and to project people and things elsewhere in time and space. Thus, the "annihilation of time and space" became a pervasive trope for those who chronicled these remarkable changes, and just as Muybridge's photographs of animal motion were said to stop time in its tracks, the Lick's giant telescope was heralded as a "mighty annihilator of space."[55]

Heroic exploration of the unknown pervaded the narratives used to describe such enterprises, never more so than in the case of the Lick, which was to be the world's first permanent high-altitude mountain observatory. As Lane has noted, it was in this period that the geographical authority of high-altitude western landscapes combined with the instrumental authority of remote sites to make mountains both special and necessary places for astronomy.[56] With these changes emerged a new type of practitioner, the astronomer-adventurer, a distinctively western identity steeped in the mythology of the American frontier. As Frederick Jackson Turner would capture in his remarkable 1893 essay, "The Significance of the Frontier in American History," for advocates of the West, the frontier was the defining crucible of the United States, its rugged challenges "furnish[ing] the forces dominating American character." The manly independence it engendered stood in direct contrast with the Old World, the advancing frontier propagating "a steady movement away from the influence of Europe, a steady growth of independence on American lines."[57] Indicative was the *Examiner's* response to Proctor's attack on Holden and the Lick, which included an editorial that praised the "candid straightforward manner of the California astronomer" as "creditable throughout to the manliness of the American character." "The absurdity of Professor Proctor's ideas is shown so conclusively by Professor Holden," they declared, "that any apprehensions that may have been excited *by the Eastern scientist's opinion* will now be laid to rest."[58] For the California press, then, Proctor served as an archetypical "other," his British origins and East Coast residency melded into the antithesis of the West's own pioneer astronomers.

The authority of this distinctive western science derived from the Pacific slope's unique blend of environmentally advantageous terrain and expeditionary scientific culture. The idea of a mountain site was first suggested to James Lick by the grandee of California astronomy, George Davidson, a pioneering explorer and surveyor of the West Coast who had experimented with telescopes while climbing in the High Sierras. The site eventually chosen for Lick's observatory, the peak of Mount Hamilton, combined the rarefied air that Davidson had experienced on his explorations with the advantage that it could been seen from James Lick's own giant flour mill in Alviso, Santa Clara County.[59] Once established as the observatory's director, Holden would glorify the site with a monograph on *Mountain Observatories in America and Europe* (1896), a foundational document for the new discipline of high-altitude astronomy that would anoint California's Lick as its inception point. Images and descriptions that stressed the rugged isolation of such mountaintop observatories became a powerful and oft-used trope for promoting the legitimacy and special status of these new institutions. By creating a strong visual connection between observatories and western mountain landscapes, and by emphasizing the climatic challenges and isolation that their astronomers had to over-come, such images tapped into powerful popular reverence for sublime mountains and frontier manliness. Like Muybridge's panorama pho-tographs or Buffalo Bill's Wild West shows, the Lick Observatory and the mythology that bolstered it therefore rested upon the technological West's unique synthesis of frontier heroism and high technoscientific industry.[60]

The greatest achievement of this synthesis might be characterized as a transformation in *seeing*. Just as Muybridge's work revealed "what had always been present but never seen," so Lick's ambition was to harness the unique, exquisite conditions of the California mountains to see deeper into space than any European or East Coast institution.[61] The media promotion of such a project as uniquely heroic and distinct-ly western was an integral facet of its power. The *Daily Alta California* would declare Lick's project "The Greatest Scientific Work of Ameri-ca," while the *Examiner* would use Proctor's attack on the Lick to press home the exacting geographical distinction that now underpinned the authority of their local institution: "A great telescope like [the Lick's], which is the pride of California and which has elicited the admiration of scientific men in all quarters of the globe, will not fail because a disgruntled gentleman of much learning and bad temper concludes to disparage it."[62] The astronomical discourse of the technological West thus shifted decisively toward the lionization of astronomical technolo-gy itself, and the purity and isolation of its exquisite location.

Exemplary of these trends is the establishment of the Yerkes Ob-servatory, another astronomical site of the technological West funded entirely by a robber-baron philanthropist, the Chicago railway mag-nate Charles Yerkes. Intended by Yerkes to "lick the Lick," the Uni-versity of Chicago's new observatory was equipped with a telescope possessing an aperture a crucial four inches larger than its Californian rival. As William Cronon has noted, late nineteenth-century Chicago was the gateway to the United States' "Great West," and that city's brash surpassing of the Lick speaks to its growing wealth and confi-dence, as well as to the escalating competition among those who had grown rich exploiting the expansive western territories.[63] The tele-scope's giant mounting was displayed at the 1893 World's Columbian Exposition (figure 2.2), and attendees of the first World Congress of Astronomy and Astrophysics held in the United States gathered at its base to hear speeches by, among others, the maker of the telescope's giant forty-inch lens, Alvan G. Clark. Speaking on "Great Telescopes of the Future," Clark recast the entire history of astronomy in terms

Figure 2.2. The Yerkes forty-inch telescope mounting on display at the 1893 Columbian Exposition, Chicago. This photograph was the celebratory frontispiece to the October 1893 issue of the American journal *Astronomy and Astro-physics* (vol. 12, plate 35).

of improvements in the construction of refracting telescopes, telling his audience that "most of the important original discoveries in the truly visual line have been made with the largest telescopes in use at the time."[64] World's fairs and their attendant scientific congresses therefore played an increasingly important role in promoting the cutting-edge technoscience that began to distinguish American astronomy after 1880. As public sites of display, they offered an extraordinary cultural synthesis of industry, consumerism, and empire. Never more so than in the case of Chicago, they also bolstered an emerging national confidence in American science and commerce that had been driven by the technological West.[65]

NEW MESSAGES FROM MARS

Proctor's own work began to look entirely out of place next to this new configuration of the new astronomy. Attention to genre is again revealing. When, at the end of 1887, Proctor published his last and most remarkable theoretical synthesis, a wide-ranging and extraordinarily ambitious essay on "Varied Life in Other Worlds," he did so in a modish Chicago journal "devoted to the work of establishing ethics and religion upon a scientific basis."[66] Proctor's ideology of astronomical practice, which relied upon the moral force of far-reaching and synthetic theorizing, began to look less like astronomy and more like philosophy or even scientific theology. The nebular hypothesis, a geological timeframe, Darwinian evolution, physical cosmology, religious doctrines of plurality, and spectrochemistry all intermingled in Proctor's essay, a constellation of disparate themes melded together to underpin broad theories about the birth and death of planetary life.

The irony is that it was Proctor's intervention against the Lick that brought the dissemblance of his approach into sharp focus. By directing attention toward the supposed failure of new, remote, specific sites of astronomical practice, Proctor had thrown into clear relief exactly what it was that made these sites special. As a weapon against Proctor's universality, certain astronomers began to stress individuality. And with the transatlantic news media gripped by the profound power of the technological West, their coverage and therefore astronomy's discourse shifted in favor of the tantalizing potential of the unique penetrating power of a small number of high-altitude observatories. By the time of the 1892 perihelic opposition of Mars, the London correspondent of the *New York Times* judged it entirely reasonable "to see how universally precedence has been given the news from the Lick Observatory over all others, and how all take it for granted that the most valuable results of observation come from there."[67] It was no longer sufficient to assimilate, speculate, and hypothesize—one had to *see*. The success of mountain astronomy hinged, therefore, on securing the claim that only the marriage of cutting-edge optical technology to the rarefied

air of remote sites and the disinterested scrutiny of the rational astronomer could produce the exquisite gaze necessary for such seeing. In a
modernized world in which, as Solnit puts it, "human beings were no
longer contained within nature," Proctor's communal, universal, and
natural argument was trumped by a technological one. Thus, as astronomy moved from metropolitan centers to isolated mountain peaks, it
also changed from a placeless, moral enterprise into a location-specific,
high-cost, technological one.[68]

While these transformations were highly localized in terms of
their sites, their impact encompassed the entire transatlantic world,
and frequently beyond. Remote yet networked, the sites of the technological West were incorporated into Anglo-American mass media from
their inception. In this sense, the Lick's telescope annihilated space in
two complementary ways: its giant lens penetrated further into the
solar system than its humbler predecessors, and its news from worlds
elsewhere also traveled further and faster than traditional mediums
of astronomical discourse. In this way, the technological West's acts of
seeing bridged the new localism of astronomical technology with a renewed global fascination with the red planet. Time mattered as much
as space here because news cycles in this new visual order were bound
together with the astronomical cycle of Martian oppositions. Mars had
been hard to observe for much of the 1880s due to the poor quality of
the planet's biennial closest approaches, a problem of limited consequence to Proctorian accounts but decisive for visual reports. As the
decade drew to a close, the coming perihelic opposition of 1892 promised the best seeing conditions since the epochal events of 1877. Proctor's sudden and unexpected death from yellow fever in 1888 marked,
therefore, a turning point that he almost certainly could not have re-
sisted had he lived. A very few new massive telescopes now offered the
potential to definitively answer the question of life on Earth's rapidly
approaching neighbor.

With this transformation came new messages from Mars. With
"seeing" the new entry requirement for news from the planet, popular
interest on both sides of the Atlantic began to look beyond Proctor's
philosophically diverse *imaginative* theoretical accounts in favor of two
concrete *visual* facets of the planet that were best seen with giant telescopes. These accounts tellingly mirrored the technologies that made
them possible. As new global, near-instantaneous communication systems began to link the West's exquisite sites to international audiences,
news from Mars turned concurrently to talk of two such technological

systems: on the one hand, a massive transnational infrastructure network visible there and, on the other, space-annihilating messages that might pass rapidly between the optically privileged observers on both worlds. Talk of Mars turned from varied life and Earthlike habitats to *canals* and *signals*. This was not, it must be stressed, a simple shift from philosophical speculation to technological objectivity. The new tools of the technological West may have promised unparalleled powers of penetration, but with them came new challenges of contested vision. As Martin Willis has persuasively argued, shifts to objectivity in the age of mechanical imaging, messaging, and recording are tempting ideas that do not bear up to historical scrutiny. Rather, seeing remained a difficult and contested practice through the turn of the century. The role of imagination changed rather than diminished, with the porous boundaries of seen and unseen shaping a kaleidoscopic range of new Martian visions.[69]

On the one hand, this renewed transatlantic fascination with the red planet naturally centered on Schiaparelli's *canali*, first sighted in 1877 and contentiously resighted, or not, under poorer observing conditions ever since. But before Mars even reached perihelic opposition, the approaching planet offered another more decisive opportunity to offer up its secrets. Might it be possible, people began to wonder, for the new power of the technological West to draw Mars directly into its rapidly expanding web of information gathering and exchange? Capturing definitively the technological triumphalism of the age, the famed French astronomer and popular author Camille Flammarion confirmed instantaneous, long-distance signaling to be the defining challenge of new Martian studies. "Everything," he declared, "leads us to believe that the planet Mars, older than the Earth in chronologic order, more quickly cooled on account of its lesser volume, more advanced in its planetary life, is at present inhabited by beings more intelligent than we, and less imperfect. But what are they? We need to be able to enter into telephotic communication with them. We do not despair of this, and we believe in progress."[70] Here, Proctorian analogies of an advanced Earthlike world are presuppositions upon which concrete technological challenges are brought to bear. Like so many fin-de-siècle contests of bold technoscientific exploration, public impetus for this work came in the form of an international prize. Flammarion had broken into the English-language popular astronomy marketplace as something of a successor to Proctor, and writing in turn in a French journal, an American Newspaper, and a British periodical, he announced a staggering hundred-thousand-franc prize left in an 1891 bequest "for the person, of no matter what nationality, who shall discover within ten years from the present time a means of communicating with a star (planet or otherwise) and of receiving a reply. The testatrix has especially in view the planet Mars, upon which the attention and investigations of *savants* have been directed already."[71] As the famed British polymath and public intellectual Francis Galton explained in a letter to the London *Times*, sending and receiving such a message was not itself a difficult task in theory so long as each correspondent had access to similarly powerful technology. With Mars "now so near us that the exceptionally large magnifying power usable at the Lick Observatory brings it optically to within 50,000 miles," mirrors could plausibly be used to flash a visual, mathematical signal using reflected sunlight, supposing only that Martians possessed a telescope of similar size.[72] Flammarion concurred, blithely assuring his readers that to message the inhabitants of Mars was, in fact, "not at all absurd, and is perhaps less bold [an idea] than that of the telephone, or the phonograph, or the photophone, or the kinetograph." Indeed, technological mastery over time and space now overwhelmed celestial distance, engulfing the moon and Mars as "celestial province[s] annexed by nature herself to our destiny." To expect reciprocal relations from our nearest planetary neighbor was merely to suppose that they too had mounted their own technological conquest of the heavens.[73]

As a cultural phenomenon, then, speculation over interplanetary communication spoke of a powerful new symbiosis between public fascination with life on Mars and massive interest in instantaneous, long-distance forms of communication—be it telegraphy, the photophone, the telephone, motion pictures, spiritualism, or newfangled "wireless" signals. Thomas Edison's first patent for a wireless telegraph system was issued in December 1891, and the famed inventor described ambitious plans to transform the iron core of his mine in Ogden, New Jersey, into a massive receiving coil capable of detecting interastral magnetism. "The electric telegraph is another marvel which has transformed the world," Flammarion wrote in response, "and may there not exist between the planetary humanities psychic lives that we do not know of yet?"[74] The topic received enough attention to draw a satire from *Punch*, which published "Extracts from the Note-book of the Secretary of the Earth and Mars Intercommunication Company, Limited." Set in a future 1927 when contact is finally made, *Punch*'s Martians are found to be too busy signaling Jupiter to bother speaking with Earth. By 1934 the frustrated Intercommunication Company is simply merged into the "London, Jupiter, Venus, Mars, and North Saturn Aerial Railway Company." Plays, cartoons, poems, songs, and a wide range of nonfiction toyed with the means and consequences of such interplanetary communication, while engineers such as Nikola Tesla discussed the practicalities of establishing a working link (see figure 2.3).[75]

As had been the case with imaginative astronomy, it was not at all clear to some practitioners whether this new technological conquest of Mars was to the benefit of their discipline. While men of science including Galton, Flammarion, and Asaph Hall took the question of signaling seriously, others, such as an anonymous editorial in the *Journal of the British Astronomical Association*, lamented what they considered to be an essentially journalistic "blunder." "The leading idea in several papers," the editorial complained, "seemed to be the prospect of starting an interplanetary telegraph." These concerns explicitly linked the power and danger of a newly networked mass media to the notion of Martian correspondence: "If one may judge from the telegrams and articles which have appeared in the newspapers . . . astronomy is making great progress in popular interest, though much that has been written shows that the public still require further education. To begin with, it was evidently expected that Mars would do something extraordinary,—flash a congratulatory communication by the Morse code at least—to celebrate his coming successfully out of opposition."[76] Proctor's old sparring partner Norman Lockyer was one commentator from the astronomical elite who took particular umbrage at this new paradigm of Martian telegraphy. Writing in his own weekly journal, *Nature*, Lockyer moved to police transmission systems that he judged had exceeded disciplinary authority. "If everything that one sees in print be true," he complained, "the inhabitants of Mars are signalling to us, and it only remains for us to choose our manner of reply. *Of course from signals the imagination of the ready writer has passed at once to words*, and having got so far, each planet is about to become acquainted with the history and present conditionings of the other by means of a language understood of our neighbours as well as ourselves."[77] At stake was a vital distinction between signals— instantaneous, ambiguous—and messages—composed, authoritative. Yet it was exactly this distinction that the global cable networks so easily confounded. Lockyer's own journal was a case in point. When news of a strange light flash on Mars became a global news sensation in early August 1894, *Nature* received its message from the same source

Mars at the Telephone: "Well, well, what is it? who is it?"
Voice in Telephone: "Tesla."
Mars: "Who *on earth* is 'Tesla'?"

Figure 2.3. "Mars and the Earth," by Edmund J. Sullivan. Satires like this cartoon from *Pall Mall Magazine* (vol. 23 [March 1901]: 364) made explicit the linked potentialities of astronomy's enhanced power to see and mass media's new capacity to transmit and receive messages across vast expanses of time and space.

as any "ready writer," and was itself as ready as the daily papers to transmit equivocal and fast-moving news from Earth's near neighbor:

> Since the arrangements for circulating telegraphic information on astronomical subjects was inaugurated, Dr. Krueger, who is in charge of the Central Bureau at Kiel, certainly has not favoured his correspondents with a stranger telegram than the one which he flashed over the world on Monday afternoon:—
>
> *"Projection lumineuse dans région australe du terminateur de Mars observée par Javelle 28 Juillet 16 heures Perrotin."*
>
> This relates to an observation made at the famous Nice Observatory, of which M. Perrotin is the Director, by M. Javelle, who is already well known for his careful work. The news therefore must be accepted seriously, and, as it may be imagined, details are anxiously awaited.

What one made of such news depended upon how one read equivocal signals broadcast over vast distances at great speeds. Even as they "anxiously awaited" further details, *Nature* had no compunction about venturing some plausible "physical or human origin" for the light flash. "Without favouring the signalling idea before we know more of the observation," they hedged, "it may be stated that a better time for signalling could scarcely be chosen" given the planet's position in the night sky.[78] Whether this constituted the imagination of the ready writer or the reasonable speculations of a correspondent responding to breaking news seemed to rest on the entangled judgments of both the gatekeepers of this new astronomical information order and their diverse readers.

This made the global news economy both a crucial resource and a cautionary lesson. In this fine distinction, sound public instruction could be forged. Francis Galton, in his 1896 *Fortnightly Review* essay on "Intelligible Signals between Neighbouring Stars," opened by noting that, as the "possibility of exchanging visible signals with Mars" was now much discussed "in the newspapers," it was worth considering how such correspondence could be made *"intrinsically* intelligible." Mathematics was the necessary vernacular, and the rapid back-and-forth of the news cable was the obvious pattern of discourse. "The simplest way of explaining my method," he wrote, "is to suppose that Mars began to signal, to the wonderment of our astronomers, who sent descriptive letters to the newspapers from day to day, out of which the following imaginary extracts are taken." Mirroring Javelle's message from Nice, Galton's account opens with an excitable announcement of "minute scintillations of light" proceeding from the planet, followed by two complementary responses. First, there was journalistic anticipation of what certain observers in specific, privileged locations might discover next: "It is hoped that the Director of the X. high-level observatory, where the atmosphere is singularly transparent, which is favourably situated for now observing the spot in question, and which is furnished with a telescope peculiarly well suited for examining Mars, will soon be able to tell us something more about it." In quick succession, however, came a cautionary addendum: "In the meantime it is well not to indulge too freely in wild speculation." Galton was not wont to let his audience down, however, so—unlike in the case of Javelle, where the next phase of Martian signaling had proven unforthcoming—his subsequent reports soon confirmed "that the scintillations are purposive," and within another half-dozen back-and-forth messages a "large typed telegraphic dispatch appeared in all the evening newspapers—COMPLETE DECIPHERMENT OF THE FIRST PART OF THE MESSAGE FROM MARS."[79]

What Galton simulated here for his readers would certainly have been familiar, even if the ultimate results were not. His account cleverly exploited the inherently serial form that telegraphic astronomical news took, highlighting journalism's now vital intermediary role as both receiver and transmitter of information.[80] Of particular significance for readers in the era of Martian signaling was the change in the speed of this news. This greatly compacted the serial progress of sending and receiving, and Galton both aped and extended the way in which recent reports of Martian signals had used anticipation and caution concurrently, a hedged response to immediate reports that were both factual and uncertain. What Galton's dramatization shows is that the solution to this limited and uncertain vision was to be found not in Proctorian imaginative astronomy but rather in the rapid concatenation of privileged astronomical reports from a few remote observatories. These alone held the power to move from uncertain signals to certain messages.

The role of imagination was both highly specific and highly contained within this new information order. Its purpose was not to extend interpretation beyond the limits of vision but rather to build scientific narrative into serial information exchange. Its use was therefore a delicate matter. Even as he deployed the familiar model of newspaper reportage to frame his imagined account, Galton was careful to preface it with a clear explanation of its fictional originals during the course of a "somewhat dreamy vacation." Galton's editor evidently favored a less expository structure and upon receiving the first draft immediately asked if the author would "mind if I omitted the opening sentences and made the paper read like a genuine account, only disclosing at the end that it was imaginary?" This suggestion Galton rejected outright, lest the piece be misread.[81] For Galton, then, the role of imagination was clear: it was for reasoned and explicit narrative instruction, not flighty projects of speculation and moral improvement.

But Galton's model was only one ideal. Historians have recently noted that repeated articulations of just how uncertain astronomical vision was in this era opened up a gap for new kinds of testimony and new genres of Martian scrutiny.[82] No author exploited these opportunities with more prowess and success, of course, than the precocious young dropout from South Kensington's Normal School of Science, H. G. Wells. As both a trained scientist and a writer keen to make a living from a diverse range of popular authorship, including fiction, Wells occupied an unusual position in late nineteenth-century intellectual culture. As influenced by Jonathan Swift's utopian satire as he was by his beloved teacher Thomas Henry Huxley, the young Wells spent a considerable amount of time contemplating the relations of actual and imagined in an era of global communication. One outlet for his early musings on these topics was Lockyer's journal, *Nature*, where in 1894—coincidentally just a week before the news of Javelle's light flash from Mars—he published a short commentary on the decline in standards among those who "popularized" science. Tellingly, Wells's critique hinged not on the moral ambitions or philosophical depth of such works but rather on the scope and limitations of serialized, narrative discourse: "The taste for good inductive reading is very widely diffused; there is a keen pleasure in seeing a previously unexpected generalisation skilfully developed. . . . The fundamental principles of construction that underlie such stories as Poe's 'Murders in the Rue Morgue,' or Conan Doyle's 'Sherlock Holmes' series, are precisely those that should guide a scientific writer. These stories show that the public delights in the ingenious unravelling of evidence. . . . First the problem, then the gradual piecing together of the solution."[83] Poe and Doyle were more than just analogical models for Wells. It was in their own blending of fact and fiction, through a serialized, sequential account, that the power of their scientific exposition lay. So, *pace* Galton, Wells's own articulations of this narrative ideal came from precisely *not* separating out the imagined from the real. Wells studied science and wrote about it as a journalist right through his career as a novelist, and he did not see fit to sharply distinguish the two modes of discourse. "I am trying," he wrote to his colleague Grant Allen in 1895, "to cultivate this field of scientific romance with a philosophical element."[84]

Mars was, of course, the most fertile ground for this form of cross-genre analysis. As a journalist, the young Wells wrote about Javelle's light flash observations and their implications, anonymously authoring "Intelligence on Mars" for the *Saturday Review* in April 1896. A year later, his first Martian romance, "The Crystal Egg," dramatized the consequences of instantaneous interplanetary communication facilitated via a series of visual portals—the eponymous eggs. The human discoverers of this remarkable two-way connection lose their egg before its full capacities can be explored, leaving the narrator to conjecture that "the terrestrial crystal must have been—possibly at some remote date—sent hither from that planet, in order to give the Martians a near view of our affairs."[85] Within six months, Wells would explore the full implications of this Martian surveillance in an extended serial for *Pearson's Magazine*, published shortly afterward as the novel *The War of the Worlds*. This full-blown invasion fantasy has been read as a great many things, including a critique of British colonialism, an attack on organized religion, an intervention into debates over Darwinism, a pessimistic warning about scientific hubris and future warfare, a wry love-letter to Woking and the home counties, and of course as pioneering science fiction.[86] Little, however, has been made of Wells's own immersion in the Martian signaling debates then raging as he typed the story for *Pearson's*. Like his factual *Saturday Review* piece before it, which had begun with the declaration that the latest "outcrop of speculations" over life on Mars had grown "from the discovery by M. Javelle of a luminous projection on the southern edge of the planet," Wells began his fictional take with the same ambiguous signals: "The storm burst upon us six years ago now. As Mars approached opposition, Lavelle of Java set the wires of the astronomical exchange palpitating with the amazing intelligence of a huge outbreak of incandescent gas upon the planet. It had occurred towards midnight of the 12th, and the spectroscope, to which he had at once resorted, indicated a mass of flaming gas, chiefly hydrogen, moving with an enormous velocity towards this earth."[87]

Wells was obviously relying here upon his readers' familiarity with these barely concealed allusions. But he was also using fiction to explore and interrogate recent news from Mars. *The War of the Worlds* may be about many things, but it is certainly concerned with the scope and limitations of global and interplanetary communications networks and with the hopes and threats that these networks presaged. Written as it was during the winter of 1895 and summer of 1896, this meant yoking the invasion of an intelligent race of Martians to the serial progress of telegraphic news distribution. After it becomes apparent in the novel that the object fallen on Horsell Common probably contains living creatures, the journalist friend of the narrator rushes to the nearest train station "in order to telegraph the news to London." "The newspaper articles," the narrator then explains, "had prepared men's minds for the reception of the idea." Aping the morning and afternoon print cycle of metropolitan papers—and so too the style of Poe and Doyle—Wells unveils his story piece by piece, escalating the threat as the news intensifies. "The early editions of the evening papers had startled London with enormous headlines: 'A MESSAGE RECEIVED FROM MARS, REMARKABLE STORY FROM WOKING,' and so forth." Naturally enough, this serial progression by necessity runs out of and through one core network of nodal points: the astrophysical observatory. "Ogilvy's wire to the Astronomical Exchange," the narrator explains, "had roused every observatory in the three kingdoms." The Martians, in response, wage a campaign of demoralization and destruction, including "hamstr[inging] mankind" by cutting "every telegraph." Ultimately, however, the Martians' mastery of long-range communications and transport technologies proves to be hubristic and short lived, undermined as it is by their bodies' inability to cope with a hostile alien environment.[88]

As a work of popular science, *The War of the Worlds* took its place among a wide variety of pieces across a range of genres that distributed, analyzed, and commentated upon debates over Martian signals and the question of intelligent life on Earth's outer neighbor. Understood in this context, it is unsurprising to find Wells's story examined in a *Review of Reviews* roundup of the latest news from Mars, sandwiched between an account of Flammarion's "discussion of the old question, 'Is Mars Inhabited?'" and an extract from W. T. Stead's spiritualist journal *Borderland* that "professes to be a description of Mars communicated by a Martian through a medium."[89] Common to all three works was a use of narrative that both incorporated and was shaped by the communications systems that had helped forge the technological West, and which now encompassed the globe.

Signals—between planets and across media genres newly interconnected by massive technological systems—sparked a kaleidoscopic fragmentation and recombination of Martian news. At the heart of this new information order were the space-annihilating tools of the technological West. As observatories moved to remote peaks connected by global telegraphy, the far-reaching speculation of imaginative astronomy was itself annihilated by the speed and reach of a system that prioritized seeing over projection. The stories one got from such an enterprise changed too, as did the social conditions of access for those who aspired to chronicle its progress. As "seeing" became news, it was the high technology and bold exploration inherent to remote astronomical sites that defined the entry requirements for the next generation of astrophysical news. And because the few observatories that met these entry requirements were located out of sight, their ability to function as news distribution centers would be a central facet of this new high-altitude astronomy. At no time would this be clearer than in 1892, when Mars moved into its first perihelic opposition since the epochal discoveries of 1877.

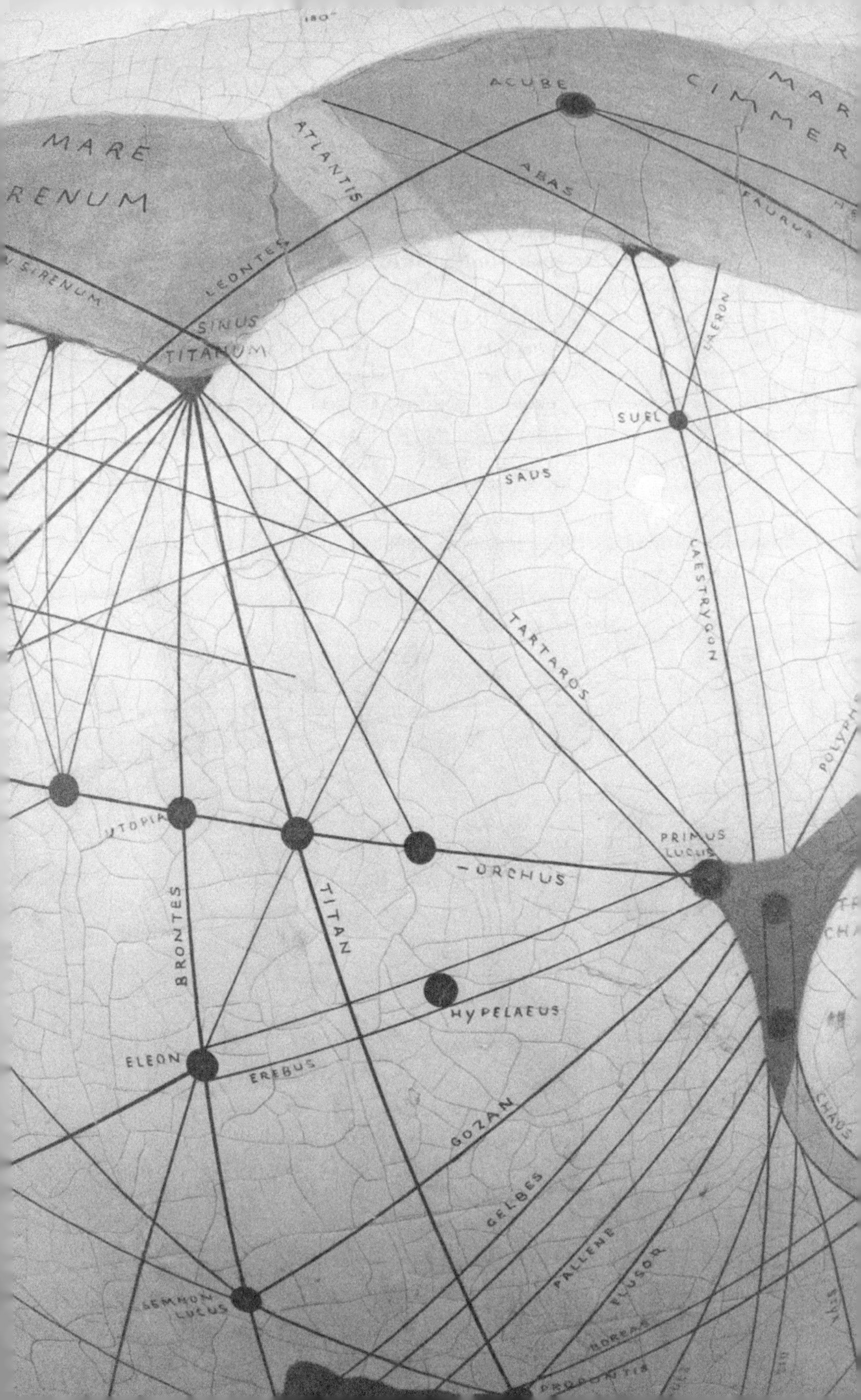

180°
ACUBE
MARE
ATLANTIS
ABAS
LEONTES
SINUS
TITANUM
SUEL
SAUS
LAESTRYGON
TARTAROS
UTOPIA
ORCHUS
PRIMUS
LUCUS
TITAN
BRONTES
HYPELAEUS
ELEON
EREBUS
GOZAN
GELBES
PALLENE
BOREAS

CONSTRUCTING CANALS ON MARS 3

Event Astronomy and the "Great Mars Boom" of 1892

A lady of the inanely inquisitive kind having met an eminent astronomer, implored permission to ask him one question. "Certainly,
madam," he replied, "if it isn't about Mars." It was about Mars. That
was the time of the great Mars boom, when public imbecility and
journalistic enterprise combined to flood the papers and society with
"news from Mars," and queries concerning Mars, most exasperating
to grave thinkers and hard workers in science. The occasion of the
excitement was this. On August 4, 1892, Mars stood right opposite to
the sun, at a distance from the earth of less than 35,000,000 miles.

—*Agnes Mary Clerke, 1896*

Four years after the events of 1892, Agnes Mary Clerke looked back
on the "great Mars boom" with a mixture of annoyance and curiosity.
As she knew well, this episode had reinvigorated and amplified the
era's most controversial and public astronomical debate, over whether visual and spectroscopic evidence indicated the presence of life on
Earth's outer neighbor. Water had been detected on the planet, seasonal changes to its appearance had been linked to a vibrant Earthlike
climate, and its curious canals had been sighted and resighted. Writing
for the *Edinburgh Review*, Clerke assessed the latest slew of works to
appear on these subjects, taking the opportunity to look back at the
episode and expose what she saw as its unfortunate consequences.[1] As

the era's most widely respected historian and analyst of astronomical work, Clerke spoke explicitly for the discipline's establishment base, its "grave thinkers and hard workers," many of whom doubted the more sensational conclusions that had been drawn during and immediately following the drama of August 1892. Tellingly, Clerke's critique linked the exasperation of her astronomical colleagues to their work's entanglement with outside agents, above all an inquisitive but slow-witted public and a cabal of unscrupulous journalists who stoked and manipulated that public's interest. For its participants, then, the great Mars boom was an astronomical event inextricably bound up with issues of communication and reception. Public fascination with the possibility that Mars might be inhabited was a component of, rather than a secondary consequence from, astronomical work on the planet.

The great Mars boom is therefore a particularly fruitful case study for extending this book's interest in what James Secord has dubbed "knowledge in transit." Secord's manifesto joins a wider body of scholarship that takes seriously the material, temporal, and spatial dimensions of the texts, images, and objects that constitute the doing of science. These works are, by and large, accounts of authorship and reception; they take authorial voice, publication genre, market forces, and reader response as integral points of analysis within accounts of the scientific world. Their focus, therefore, is usually on books and journals, and the trajectory they trace is typically very clear, from the individualized sites of knowledge production out into the marketplace of the sciences' various publics.[2]

This chapter aims to augment such trajectories by shifting our focus away from reception and toward transmission. In what follows, public forms of discourse are shown to have been fundamentally embedded *within* the astronomical work that made the great Mars boom. To see that shift, we must turn our center of attention toward the material movement of information during the processes that constitute scientific labor itself. A central focus here, therefore, is the material dimensions of transmission technologies, considered both in terms of specific material tools (e.g., the telegraph), as well as more general media genres (e.g., the newspaper). Such technologies, designed to enable knowledge to travel, also impose specific limits on how that knowledge can travel, shaping and bounding the movement of information. This means that they impact the scale, both spatial and temporal, on which the sciences are able to work. Furthermore, their use is neither neutral nor passive, requiring considered choices on the part of the actors who seek to deploy them and constant work to make those choices count. Despite this work, communications technologies almost always produce effects that exceed or undermine the intended functions that those actors hoped they would serve. These are practical problems that impinge upon and thus in part construct scientific work itself. More than mere conduits, communication technologies are embedded within the working worlds of the sciences—multiplying, stabilizing, and therefore co-constituting the knowledge these worlds produce.[3] This means that as well as thinking of knowledge in transit as a matter of authors and audiences, we also need to think of it as a practical entanglement of scientific praxis and material transmission technologies. By pushing our focus back into the laboratory or observatory, and then out from there via material means—wires, submarine cables, newsprint—we can observe the effects that these technologies have on shaping, and therefore making, knowledge.

EVENT ASTRONOMY AND TRANSNATIONAL TELEGRAPHIC NEWS

First we must return to the technological West and think more carefully about exactly how the new remote astronomical sites described in the previous chapter distributed novel observational claims in an age of telegraphic media. We have seen in general terms how astrophysics' reorientation in space occasioned a concomitant reorientation in its relations with mass media. In this section I argue that one significant consequence of these changes was an entirely new form of mediatized work, which I call "event astronomy." What I hope to show are the ways in which the prioritization of seeing and the dramatic reduction in institutions and practitioners capable of seeing generated a new symbiotic relationship between particular astronomers and particular media agents. I first characterize this relationship by looking at an eclipse expedition lead by Harvard astronomers in 1889. This establishes for us the major actors, novel practices, and particular media technologies that together constitute the principal subject of this chapter: the great Mars boom of 1892.

Common to both the 1889 eclipse and the subsequent furor over Mars three and a half years later was the fact that these were *events* in the literal sense. They were singular, significant occurrences that took place at a particular time and place. These events' power—their ability to be perceived as such on a global scale—derived from the paradigm of the technological West. The practices of mountain astrophysics isolated astronomical work at specific places and times; the material technologies of the telegraphically connected news media facilitated the near-instantaneous transmission of this work internationally. What made these new enterprises newsworthy extended well beyond the moment of observation, however. As we have seen in chapter 2, by emphasizing the isolation and purity of the remote mountains on which they worked, mountain astrophysicists actively cultivated a novel persona: the astronomer-adventurer. Such a persona played into—and therefore was co-constructed by—the prevailing news economy of the era. Highlighting the rugged and demanding character of their mountain work distinguished this new breed from metropolitan colleagues, and it tapped into existing popular enthusiasm for accounts of heroic exploration. Complex, challenging, and courageous trips to observe solar eclipses and transits of Venus fit neatly into a media marketplace that voraciously consumed both astronomical news and narratives of triumphant exploration. In an era of mass readership, astronomers and the journalists who covered them "wrote for a public that liked reading about travel, adventure, cheerful success over adversity, and the exotic." Astronomers, like polar explorers, therefore sought to represent themselves as "masculine heroes, adventurous travellers, and scientific explorers," an image that the press was more than happy to promote.[4] By the 1880s competition in the United States for newspaper readers was fierce, and exclusive coverage of the new American astronomical mountain expeditions therefore joined other forms of adventure reportage as a prized technique for capturing new audiences.

One newspaper above all sought to dominate this travel and adventure market. By the 1880s, the *New York Herald* had the largest circulation of any newspaper in the United States. As the most successful product of the "penny press" revolution of 1830s New York, the *Herald* was a populist, sensationalist, progressive organ that had long prided itself on its unparalleled news-gathering capabilities. Under the stewardship of the irascible and combative James Gordon Bennett and then of his namesake son, the paper had pioneered many of the industry's technological and organizational innovations, including pony expresses, harbor news boats, lightning presses, dedicated telegraph lines, and the incorporated press agency. Above all, though, it was the paper's unrivaled network of international correspondents that set it apart from its competitors. From its earliest days, Bennett Sr. had committed his paper to an international scope, and by the 1860s it had regular and special correspondents in at least twelve countries on four continents. The *Herald* exploited the costly new technology of submarine telegraphy to gather news at lightning pace, repeatedly outmaneuvering its rivals in the fierce marketplace of international news transmission.[5] Bennett Jr. built on his father's legacy, focusing much of his paper's energies on covering, and in many instances generating, spectacular news of adventure, exploration, and scientific discovery. When he dispatched Henry Morton Stanley to Africa to find David Livingstone, Bennett Jr. is reported to have given his correspondent an unlimited budget, explaining that he intended to "publish whatever news will be interesting to the world at no matter what cost." By the 1880s adventure had become a staple of the *Herald*'s output, with the paper sponsoring polar expeditions, running competitions to promote the development of automobiles and powered flight—and offering both coverage of and tangible support to a variety of astronomical expeditions.[6]

The *Herald*'s most successful partner in this enterprise was Harvard College Observatory. Under the careful stewardship of its enterprising and exacting director, Edward Charles Pickering, Harvard Observatory had risen by the 1880s to a position of preeminence in the nascent discipline of astrophysics.[7] Pickering's ambitious program of stellar mapping, photometry, and spectroscopy was not well suited, however, to the poor seeing conditions of Cambridge, Massachusetts, and it faced the immediate threat of technological obsolescence from America's first permanent mountain site, California's Lick Observatory. Harvard's response to this western threat was the pursuit of the Boyden Fund, left in 1879 by the Massachusetts engineer and inventor Uriah Atherton Boyden to establish an observatory for "conducting astronomical observations at such a height as to be free, so far as possible, from the injurious effects due to the atmosphere." After a strenuous five-year campaign, Pickering secured the fund for Harvard in February 1887, just as the Lick neared completion. With no viable mountain location within easy reach of Cambridge, expeditions were sent west to scout potential sites for a new permanent sister observatory, with locations surveyed and tested in Colorado, Southern California, and Peru.[8] Thus began Harvard Observatory's entry into the new realm of astronomical exploration, with the *Herald* at the forefront of chronicling and promoting its ambitious plans. The astronomers sent west to climb mountains and test sites were, the *Herald* wrote, "undaunted soldiers of science," their initial results "far surpass[ing] the most sanguine expectations."[9]

Considerable promotional work by astronomers and observatory directors made this sort of positive reportage happen. Proactively nurturing relationships with the press helped raise personal profiles and—just as importantly—money for new high-cost astrophysical institutions.[10] In addition, building relationships with newspapers like the *Herald* also benefited the practical execution of scientific field trips. This was particularly evident in the case of Harvard Observatory's 1889 eclipse expedition, which was directed by the same astronomer in charge of finding a site for the Boyden station, Edward Pickering's younger brother, William.[11] A keen mountaineer, William was another archetype of that new breed, the astronomer-adventurer. At once field scientists, precision observers, and adventurer-geographers, such men had to manage a blurred identity as well as organize an array of complex travel and scientific pursuits. Collaboration with the press not only bolstered this nascent identity formation but also placed important logistical and technical assistance at these astronomers' disposal. The eclipse of January 1, 1889 proved to be exemplary of this symbiosis, helping forge a relationship between William Pickering and the *Herald* that would have a profound impact on his subsequent work in 1892.

In a stroke of luck for Harvard Observatory, the 1889 eclipse placed northern California in the path of totality. William would lead a dual purpose expedition: first, his party would observe the eclipse in the north of the state; then, they would move south to Wilson's Peak to scout the mountain as a potential site for the Boyden station. Although both Pickering brothers shared a keen eye for publicity, in the adventures that followed William would prove to be much more of a showman adventurer than his more sober brother, Edward. Eschewing the typical practice of engaging a press officer to travel with them and report on the party's behalf, William instead devised a bold first in eclipse observation reporting: direct, first-person, real-time telegraphic broadcasting from the site of observation itself. Engaging Western Union and the *Herald* as partners, William organized a dedicated telegraph line that ran three thousand miles uninterrupted from his temporary observing station in Willows, California, directly into the *Herald*'s office in downtown Manhattan. The newspaper, which underwrote this high-tech enterprise, would capture in return an exclusive report from the expedition's remote site.

Through this arrangement William secured in return both personal publicity and access to a dedicated time signal, which David Todd, director of Amherst College Observatory, signaled directly to him from the *Herald*'s New York office using a prearranged cipher. The *Herald*, by return message, printed up William's reports of the eclipse as telegraphic transcripts, complete with datelines that included the precise time of transmission, down to the minute. As was typical to the genre, the resultant reports read like explorers' dispatches, breathlessly describing the eclipse team's visual, photographic, and photometric observations of a corona that "extend[ed] outwards from the sun . . . two millions of miles." More specifically, however, they also establishing a style that would become typical of event astronomy, in which the adventurous tone of the dispatches was tempered by a cool scientific rigor that artfully drew attention to the technoscientific systems that constituted the event as a unique, difficult, and special enterprise. William went out of his way to note the scientific advantage of the telegraphic time signal, in particular, which had enabled him to secure the exact times of geometric contact and to determine the duration of totality as 118 seconds, "three seconds longer than predicted."[12] Although astronomers from numerous observatories had traveled to view the eclipse, William Pickering's unique exploitation of the transmission technologies provided by the *Herald*—dedicated telegraph lines, rapid print and distribution networks, syndicated replication of the dispatches across most U.S. cities, same-day reproduction in the *Herald*'s Paris edition—secured him his own exclusive, as the first astronomer to present his observations and discoveries in a public forum.

The impact of this collaboration endured long after the eclipse itself. In the following weeks, the *Herald* spent as much space crowing about its exclusive setup as it had reporting on the event itself. "The combination of newspaper energy, astronomical learning and telegraphic speed was never better shown than in the *Herald*'s plans for reporting this great scientific event," the paper boasted.[13] Commendations rounded up from half a dozen eminent men of science and ostentatiously printed in full also reiterated the singularity of the paper's achievement: Simon Newcomb agreed that "the *Herald* had done a wonderful service to science"; Asaph Hall declared that "the *Herald* has set an admirable example in collecting and publishing so promptly reports of the eclipse"; George Barker admitted that "all his information about the eclipse had been derived from the *Herald*."[14] As with reports of crime or society scandal, exclusives burnished a paper's reputation as preeminent in the art and craft of news gathering. Particular to this form of astronomical event, however, was the *Herald*'s self-promoting claim that its unique coverage also made it an active agent in *discovery*. David Todd's lively first-person account of the enterprise for the *Century* magazine, entitled "How Man's Messenger Outran the Moon," included an illustration allegorically depicting the team's scientific news triumph: the line of totality sweeping across the North American continent, overshadowing literally, but not figuratively, an even greater line below it, that of the *Herald's* telegraph wire straddling the entire breadth of the country (figure 3.1).[15]

Once established as mutually beneficial, these links between William Pickering and the *Herald* would endure. The two parties' next collaboration, however, would be in the more remote environment of South America. Once again, geographies of transnational media proved decisive. In terms of English-language news gathering, South America was a continent that the *Herald* completely dominated. Under the expansionist leadership of Bennett Jr., in particular, the New York paper had poured money into correspondents and telegraph networks across the continent, in a concerted effort to control its news output north into the United States and Europe. At a time of growing U.S. imperial intervention in South America, Bennett Jr. styled himself as a "sort of ex-officio, one-man state department," utilizing his correspondents to bolster the United States' "Big Sister" policy of active political and business intervention in the region.[16] At the heart of this media imperialism was the murky and complicated business of transnational submarine cables. In a move that would prove hugely significant for the great Mars boom a decade later, in 1881 James Scrymser's Central and South American Telegraph Company finally established a direct cable link between North and South America, first connecting Galveston, Texas, to Mexico City, then reaching Callao, Peru, a year later, and finally in 1891 extending a fledgling network south of Lima as far as Valparaíso, Chile, and across to Argentina.[17] Bennett Jr. immediately negotiated an exclusive news transmission deal on Scrymser's cables, effectively cornering the U.S. market for South American news. "To say that the governments and press, as well as the reading public of both America and Europe have depended on [our] despatches for South American news, is simply to state a fact universally known and recognized," the *Herald* gloated.[18]

This monopoly extended to astronomical expeditions too. When William Pickering led another eclipse expedition, this time to a remote mountain site in Chile, the *Herald's* local correspondent, Henry Wolfe, assisted with travel arrangements and accompanied the party. The

Figure 3.1. The 1889 eclipse, as illustrated in David Todd's account for *Century* magazine (vol. 38 [August 1889]: 606). Todd's article began, "It came about on this wise—rather complexly. Sun and moon, types and wires,—astronomy, journalism, and telegraphy,—all were concerned in the contest."

New York and Paris *Herald* reported the eclipse observations of William's party in detail the morning after the event, a news distribution coup that the French Academy of Sciences dubbed "unparalleled." Edward, back in Cambridge, when pressed for comment declared himself "amazed and delighted at the feat of the *Herald*"; without the budget to communicate with his brother via telegraph, he had been left otherwise dependent upon "the slow course of the mails." "The *Herald*," boasted the paper, "is always to be depended on to come to the rescue of impoverished science."[19] On his return from the expedition, William went via the paper's office in New York to stop in on journalists who were now also friends. "We spoke of you," William wrote to Wolfe, "and they seemed much satisfied to have got out the eclipse news in the Paris edition a day ahead of the European parties. I told them that was all due to your being with us, and that you had rushed things, and that I thought you were doing great things down there."[20] Harvard's remote astronomical sites thus entered into a symbiotic relationship with the internationally networked *Herald*, establishing a new power dynamic that underpinned a new form of event astronomy.

REPORTING WILLIAM PICKERING'S
1892 OBSERVATIONS OF MARS

At no time was this close relationship between astronomy, journalism, and telegraphy more spectacularly evident than in the middle months of 1892, when William Pickering's observations of Mars from Harvard's newly established South American station became the central moment of the great Mars boom. After a thorough and intrepid search across two continents, Harvard Observatory chose the remote mountain site of Arequipa, Peru, for its Boyden Fund outstation. Following the success of the 1889 eclipse field trip, Edward Pickering chose his younger brother William as the man to establish the new observatory. The *Herald*, naturally, was on hand to cover the new mission, trumpeting it as "probably the best equipped" and "the most comprehensive scientific expedition ever sent forth." These reports characterized the new site as the epitome of the technological West, singling out for praise the mountain observatory's battery of "fine instruments" and Arequipa's "transparent Andean skies."[21]

As historians have chronicled, once William arrived in January 1891 to establish the observatory, things rapidly got out of hand. Edward's conservative plans for a relatively cheap, modest outpost were immediately ignored by his headstrong and ambitious brother, who, having been sent out with instructions to initially spend no more than five hundred dollars on the site, proceeded to buy up large tracts of land and build himself an extravagant house, quickly blowing well past his meager budget.[22] To make matters worse, William's excessive spending did not appear to be producing concrete results, and by the end of 1891 Edward had still not received any appreciable quantity of data or photographic plates from his outstation. Edward's plan had always been for the Arequipa site to serve Harvard Observatory's long-term and ambitious project of systematic photometry and spectroscopy, which relied upon a steady stream of plates being shipped back to Cambridge for careful analysis.[23] William, however, had entirely different intentions. Rather than undertaking the laborious and unglamorous survey work with which he had been charged, he decided instead to concentrate on visual observation of the planets. Mars was about to come into perihelic opposition for the first time since the epochal events of 1877, and, as Edward perhaps should have known, this planet had long held a particular fascination for his brother.

With his flair for the extravagant and newsworthy, it was natural enough that William gravitated to Mars. Since the late 1880s at least, the younger Pickering had focused more and more attention on the enigmatic planet, with his burgeoning publication output on its curious physical constitution feeding on the ever-growing public interest in the subject. As he noted in an 1888 letter to *Science*: "There has been so much said of late, in the newspapers and elsewhere in regard to the parallel canals of Mars," that the subject begged further analysis. William's own opinion at this stage was that the canals were likely not waterways but vegetation, and he followed Richard Proctor in surmising that spectroscopic evidence implied the likely evolution of animal life on the planet. Observational data was still lacking, however, and in 1890 William began his own visual observations of Mars from Cambridge.[24] The *Herald* naturally took an active interest, with its international scope again proving significant. Starting in August 1890, the newspaper began publishing, through its Paris edition, regular pieces on both sides of the Atlantic by Camille Flammarion. As we have already seen, Flammarion was a committed populist with a particular interest in life on Mars, but this was likely the first time that many *Herald* readers would have encountered his work. Though his first book, *La Pluralité des Mondes Habités* (1862), was already in its thirty-third edition by 1890, his work at this point was only just beginning to be translated into English.[25] Little wonder that the *Herald*'s Paris edition was at the forefront of this transition—Flammarion's bold and conjectural style fit the paper perfectly. In a September 16 piece entitled "NEWS FROM THE PLANET MARS," the *Herald* gleefully reported that the French astronomer's recent work in the paper "strongly hints that his new observations indicate the agency of intelligent creatures" on the planet. No doubt to the *Herald's* delight, this launched Flammarion straight into a confrontation with the director of the newly established Lick Observatory, Edward Holden, who characteristically urged extreme caution when it came to drawing conclusions on the nature and significance of what he pointedly called "Schiaparelli's canals."[26]

Such debates reignited the long-simmering tensions between sensation-seeking news outlets and those practitioners of astrophysics who wished to protect the rigor and sobriety of their nascent discipline. Media choices mattered a great deal in these conflicts. Holden, in his response to Flammarion, grumpily telegraphed the *Herald* to make clear his distaste at being drawn into a mass media forum: "The details of scientific evidence cannot be discussed in the columns of a daily newspaper," he complained, as "the proper place for such a discussion is in an astronomical journal before a jury of experts."[27] But such firm statements belied the ambiguity inherent to the publication of expert astrophysical testimony in an age of mass media. William Pickering, when approached by the *Herald's* Boston correspondent for comment on the affair, quite happily characterized his own role as an expert as commensurate with vibrant speculation in public forums. This only stoked the sensation further, not least by employing a Flammarion-esque degree of ambiguity and conjecture in intimating that he saw Mars as an aged, dying planet that was (in the *Herald's* words) "largely uninhabitable." Such enigmatic statements, when printed under the sensational headline "THE PLANET MARS IS NIGH TO DEATH," will doubt-lessly have consternated Holden (and Edward Pickering too), but they also cast William as a central character in the growing saga over the question of life on the planet.[28] By the time he sailed for Peru in December, the *Herald* was excitedly proclaiming that "perhaps the most interesting discoveries to reward Professor Pickering will be the nature of the canals of Mars when the ruddy planet in 1892 makes his nearest approach to the earth."[29]

Once settled in Arequipa, William appeared to concur. Ignoring his photographic assignments and repeated pleas from Edward to send back negatives, by mid-1891 William was enthusiastically extolling the superior seeing conditions at his mountain outpost, especially with respect to visual observation of the planets. Mars was approaching, and William wrote to his brother excitedly to predict big things for the coming perihelic opposition: "Here with the fine seeing, and the southern declination of the planet, I am quite sure I should see more, and if so, probably more than anybody has ever seen before. There will not be another equally good opportunity for fifteen years. I think it should not be missed."[30] William felt justified in choosing visual over photographic techniques because he believed that a skilled observer such as himself could make out more detail on Mars's surface during fleeting moments of exceptional seeing than long exposure photographs could ever reveal.[31] His unique location afforded him an opportunity to make a name for himself with new and exciting discoveries, and it's clear that he regarded this visual work on Mars as a form of exploration. To his mother he wrote: "I have been very busy upon Mars this week, I begin to feel quite at home upon the planet, and feel as if I could quite find my way about upon it anywhere. Last night I discovered a new lake that is not shown upon any of the maps. . . . It is very interesting, and I look forward to each night's work with as much interest as if I could really visit the countries I am looking at." "In the 400th year from Columbus," he wrote in another letter, "we feel as if we too were almost discovering a new country."[32]

William himself was isolated from most of the international coverage of the 1892 opposition, though he might have suspected the press excitement that was beginning to build. Interest from newspapers and periodicals began to ramp up well before the opposition itself, stoked in particular by a renewed sensation—promoted by Flammarion through the *Herald*—over the possibility of sending and receiving visual signals to and from Mars's inhabitants.[33] Meanwhile in Cambridge, Massachusetts, Edward Pickering struggled with his own problems of long-distance communication, agonizing over how to control his wayward, free-spending brother. In the spring of 1892, two months before Martian opposition, he finally made up his mind. In late May, after much commotion among the Harvard Corporation that administered the Boyden Fund, Edward wrote a long and stern letter to William explaining that he was being recalled permanently and would be replaced by Solon Bailey in early 1893. With debt piling up and little data returning to Cambridge, Edward forlornly conceded that William was likely to continue with his own personal work, rather than concentrate on generating systematic data and photographs for computers to analyze in Cambridge: "As the routine work with the 13 inch cannot advance very far this year and must be mainly done by Mr. Bailey, there is no reason why you should not make rather more visual observations than formerly proposed."[34]

William was both shocked and deeply upset by Edward's decision, and the relationship between the two brothers began to disintegrate. Greatly complicating matters was the considerable time lag inherent to communication by post. Mail took about a month to get between Cambridge and Arequipa, and the observatory could not afford to use the telegraph except for the most urgent and brief of messages.[35] This delay gave William a significant degree of freedom, which he exploited to his own advantage. As Nisbett Becker's account of Harvard Observatory's initial U.S. high-altitude expeditions stresses, the successful management of these astronomical expeditions was dependent upon an unbroken line of communication.[36] But Arequipa was much more remote in terms of communication than Colorado or California, making William much harder to manage. Worse still for Edward, whereas he could only rarely resort to the costly cablegram, the international press could use it liberally. William, well aware of the newsworthy nature of his visual work on Mars, exploited this discord in telegraphic reach to sidestep his brother's control. Rather than methodically sending material back to Cambridge for Harvard Observatory to analyze and publish, William used the news media and the telegraph to take his own observations and claims straight to an international audience of millions. Knowing full well that it would take well over a month for Edward to get back a word in reply, three weeks before opposition William wrote to his increasingly angry brother indicating his immediate intentions: "My talents don't lie in the direction of publishing big and complete catalogues. Your talents do. I think I had better do what I am best fitted for. It is better to publish a short article well, and make discoveries, than to publish long articles poorly."[37]

William's potential discoveries took on momentous proportions even before he had cabled a word back to the United States. With the newspaper press already worked up into a frenzy over what might be seen on Mars, northern observatories, in particular the Lick, suddenly dampened the mood by reporting in early August that viewing conditions in the northern hemisphere were poor. Suddenly, South American observations assumed an enhanced significance. William Christie at Greenwich flatly rebuffed an approach from the Associated Press by reminding it that his observatory was too far north, did not have a powerful enough telescope, and suffered under "thick" weather. "The best results," he stated, "ought to be expected from the observatory of Professor Pickering, of Harvard College, who is taking observations from a point at the top of the Andes in Peru."[38] For several days around the opposition, which occurred early in the morning of August 4, readers were therefore presented with a somewhat mixed and confusing account of events. Before any report from Peru had been received and with ideological and practical reasons for downplaying the opposition's importance, Edward Holden, Asaph Hall, and Edward Pickering all went out of their way to dampen expectations and limit sensation. Hall, at the Naval Observatory in Washington, told the *New York Times* that "he did not understand why there should be such general popular interest" in the opposition and that "no special phenomena have been observed and none are expected."[39] Edward Pickering, in an interview with the *Herald* at Harvard Observatory, stated matter-of-factly that "contrary to general opinion, this opposition is of less interest to astronomers than to the general public." The *Herald*, no doubt disappointed by Edward's damp squib, attempted to wring a modicum of excitement out of the interview the only way that they could: by running it under a headline that trumpeted what exciting things might be discovered by his brother in Peru.[40] There was, however, no way of sugar-coating the surly screed that Holden telegraphed to the paper from California:

> The great interest which is felt on this subject by the public is a remarkable proof of the way in which knowledge is now disseminated by the newspaper press. . . . When every newspaper has its scientific column . . . and when the popular books and articles of Proctor and Flammarion are so ready of access . . . the effect . . . appears to be that a very great number of highly intelligent persons have a living interest in a very great number of scientific questions. . . . There is, however, I think, a bad side to this intimacy, which results from the very uncritical attitude of the public to what is printed, and I have especially remarked this point in relation to the observations of Mars at the present time. The tenor of the telegrams which I have seen so far would seem to indicate that it was expected of astronomers that they should conclusively prove that Mars is inhabited by beings like ourselves. . . . It also seems to me that several popular writers are doing far more harm than good in encouraging such baseless hopes.[41]

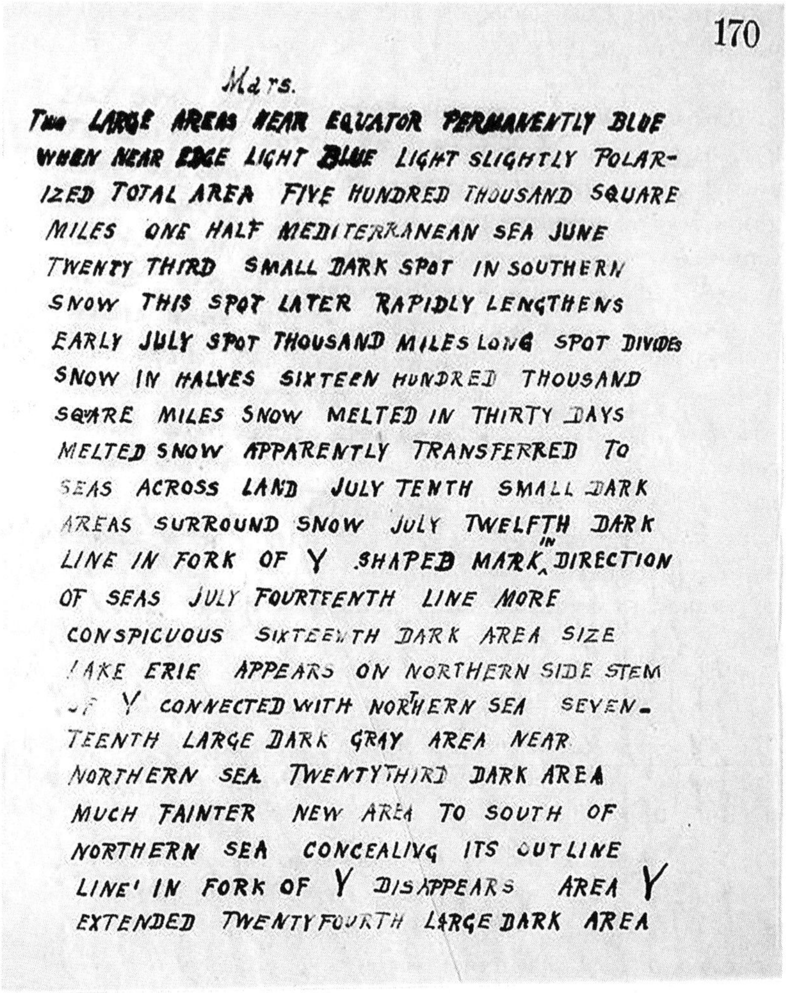

Mars.

TWO LARGE AREAS NEAR EQUATOR PERMANENTLY BLUE
WHEN NEAR EDGE LIGHT BLUE LIGHT SLIGHTLY POLAR-
IZED TOTAL AREA FIVE HUNDRED THOUSAND SQUARE
MILES ONE HALF MEDITERRANEAN SEA JUNE
TWENTY THIRD SMALL DARK SPOT IN SOUTHERN
SNOW THIS SPOT LATER RAPIDLY LENGTHENS
EARLY JULY SPOT THOUSAND MILES LONG SPOT DIVIDES
SNOW IN HALVES SIXTEEN HUNDRED THOUSAND
SQUARE MILES SNOW MELTED IN THIRTY DAYS
MELTED SNOW APPARENTLY TRANSFERRED TO
SEAS ACROSS LAND JULY TENTH SMALL DARK
AREAS SURROUND SNOW JULY TWELFTH DARK
LINE IN FORK OF Y SHAPED MARK, IN DIRECTION
OF SEAS JULY FOURTEENTH LINE MORE
CONSPICUOUS SIXTEENTH DARK AREA SIZE
LAKE ERIE APPEARS ON NORTHERN SIDE STEM
OF Y CONNECTED WITH NORTHERN SEA SEVEN-
TEENTH LARGE DARK GRAY AREA NEAR
NORTHERN SEA TWENTYTHIRD DARK AREA
MUCH FAINTER NEW AREA TO SOUTH OF
NORTHERN SEA CONCEALING ITS OUTLINE
LINE' IN FORK OF Y DISAPPEARS AREA Y
EXTENDED TWENTYFOURTH LARGE DARK AREA

Figure 3.2. Manuscript draft of William Pickering's first telegram to the *New York Herald*, sent from Arequipa, Peru, to the newspaper's Lima office on August 8, 1892. Reproduced courtesy of Harvard University Archives (HUA-ECP UAV 630.14.5, E3).

These doubts proved, initially, to be prescient. The first accounts of the opposition to be reported in the United States were from North American observatories, many of which had newspaper reporters outside their doors clamoring for news. But with bad weather and Mars's poor location in the sky having greatly diminished seeing conditions in the northern hemisphere, even the *Herald* had to admit on August 5

LIKE SEAS APPEARS NEAR MELTING SNOW
TWENTYFIFTH SOUTHERN BRANCH Y BECOMES
VERY NARROW OUTLINES NORTHERN SEA SEEN
AGAIN NARROW WHITE LINE STRETCHING
NORTH FROM SNOW MANY OTHER CHANGES
NOTED RAPIDLY CHANGING FAINT WHITISH
AREAS ~~SEEN~~ GREEN AREAS NEAR POLES
NOT SEEN FOR MANY WEEKS BUT TRACES
RECENTLY SUSPECTED AND BRIGHT GREEN
AREA DISTINCTLY SEEN NEAR NORTH POLE
LAST NIGHT

PICKERING

HERALD LIMA

that there were "NO GREAT DISCOVERIES IN REGARDS TO MARS." There remained, however, one ray of optimism amid all the gloom. "Harvard's Annex in [Peru]" explained a deck below the headline, "Expected to Furnish Much Valuable Data about the Neighbouring World."[42] All they had to do was wait for it.

The *Herald* need not fret, however, that it might be beaten to the news. Because it had exclusive access to Scrymser's newly constructed telegraphic link between South and North America, it had a monopoly on cabled news from Arequipa. With close working relations already established between the paper and William Pickering, in early August the *Herald*'s correspondents in both Lima, Peru, and Valparaíso, Chile, telegraphed the astronomer asking for two hundred words giving a full account of his observations. William, by now on bad terms with his brother and always happy to secure publicity for himself, happily obliged without first contacting Cambridge.

Harvard Observatory's archive allows us to trace the material movement of William's observations from the telescope to a global readership, all in a matter of days. First, William wrote out a 211-word message by hand, summarizing his observational notes in the typical curt diction of the telegram (figure 3.2). He then sent this message over the local Peruvian network from Arequipa to Lima, after a short delay sorting out prepayment, on August 8.[43] William's dispatch was then forwarded by the *Herald's* correspondent in Lima through Scrymser's network to the New York office, where it fell to the paper's journalists to interpret the short, unpunctuated message and transform it into a complete article. This article appeared in both the American and European editions of the paper on August 10, as well as in a host of syndicated partners, including the *Boston Daily Globe*, the *Chicago Tribune*, the *Saint Louis Post-Dispatch*, and the *San Francisco Chronicle*, all of which were able to secure local exclusives by publishing the news on the same day as the *Herald*, one full day ahead of the inevitable reprints in nonsyndicated rivals.[44]

William's account sparked just the sensation that the *Herald* had doubtlessly been hoping for. While northern hemisphere astronomers had been reticent to claim any new discoveries or draw any new conclusions about what they saw, William's account read like the chronicle of an explorer who had returned from an unknown land, teeming with new revelations. William's frenetic, clipped telegram ("EARLY JULY SPOT THOUSAND MILES LONG SPOT DIVIDES SNOW IN HALVES") was converted by the *Herald's* journalists into a breezy and exciting first-person narrative: "In my observations of Mars I have seen two large areas near the equator which are permanently blue. . . . Sixteen hundred thousand square miles of snow have melted within the last thirty days. The melted snow has apparently been transferred to the sea, across land." In a separate commentary piece, the *Herald* focused on the remarkable degree of change that William had described, the scale of which dwarfed any analogous transformations on Earth, and "perhaps thr[e]w more light upon the physical condition of a neighboring planet than have ever before been obtained."[45]

Because these observations were new discoveries, describing details and features never before reported, the *Herald* and its syndicated partners presented their articles as a discreet news event: the scoop. William's account stood out precisely because of its entirely novel observations, and the immediate effect was to reignite debates about seeing and the work of mountain observatories. Seen in terms of isolated sites with privileged views of Mars, William's report was interpreted by some newspapers as a victory for Harvard over other key northern hemisphere observatories, in particular the Lick in California. The debate that ensued was, simultaneously, about the technological priority of those few, privileged sites, the observational competence of the small number of astronomers who worked at them, and the journalistic propriety of sensation-seeking editors. The consequence of these entangled arguments was the emergence of a new narrative about Mars, centered on one single feature of the planet.

TELEGRAMS, INTERPRETATION, AND THE EMERGENCE OF THE "CANAL" NARRATIVE

Before William Pickering's telegram arrived, it had looked to some like the *Herald* might be dangerously overhyping its Mars coverage. Reliance on Flammarion's exciting cables from France, in particular, had exposed the paper to embarrassment, given that nothing new of note had so far been seen. When Holden began working the press hard in early August with his message of caution and skepticism, certain rivals of the *Herald* jumped at the opportunity to use these assertions to mock its sensation-mongering rival. Holden had explicitly rejected the exploration trope, stating that "the case is not as though astronomers were pushing out on an unknown ocean expecting to find an unknown continent or a passage to India beyond the sea. In such a case the mere announcement of the discovery would be all that was needed and expected. The actual circumstances are very different. It is as if someone had undertaken to make a new survey of the State of Arkansas, for example."[46] On August 6 the *Washington Post*, in a tongue-in-cheek editorial, dubbed the Lick's director "Heartless Holden," jokingly chastising him for "crushing" the claims of "*journalistic astronomers*" who were happily "giving to the world the fruits of their unleavened ignorance."[47] A day later the *New York Times* went even harder at Flammarion and the *Herald*, publishing a spoof report on "The Folks on Mars" that equated the French astronomer with a psychic medium in Denver and the "weather prophet" E. Stone Wiggins: "Old-fogy astronomers . . . are always averse to telling what they know quickly enough to make the information available for timely and lively articles in sensational newspapers, but happily there are other scientists always willing to tell a great deal more than they know, if necessary, in order to entertain the public."[48] This pointed criticism came crashing down three days later, when the *Herald* scooped its rivals to the freshest, most impressive news from Mars yet to appear.

With William Pickering's account as the new standard for Martian astronomy, the narrative was quickly reversed. Holden and his conservative colleagues were now singled out as failures who had flunked the astronomical challenge of the decade because of their overly cautious approach. On August 11 the *San Francisco Chronicle* reprinted at length a particularly cutting reproach from the *Brooklyn Times* that accused Holden of being an incompetent observer who had secured his position through backroom political influence and had then used it to charge "$30 a column" for the privilege of printing accounts verified by the "Holden trade-mark." His greatest crime, however, was that he had almost succeeded in ruining the astronomical news event of the season. "When the eyes of the United States were fixed on Mars, lo, there stood between the star and the people the form of Holden. He notified the expectant public, as with the voice of an oracle, that there was not much to see, anyhow."[49]

Although astronomers at the Lick received William Pickering's news from Mars "with a kind of amazement," they were clearly backed into a corner.[50] Either Holden and his staff were not competent enough to see what William saw, or their site was inferior to Harvard's outstation. For the press, the notion that William's observations might be in error did not enter the conversation. The credibility of California's preeminence in the new mountain astronomy thus hinged on both what the Lick could do next and what the newspapers would say next. Even papers that were natural enemies of the *Herald*, such as the *New York Times*, and that had lauded the preeminence of the Lick before August 10 started to give preeminence to William Pickering's observations afterward.[51] On August 18 Holden moved to preempt further debate, sending a telegram to the *Herald* syndicate's principal rival, the Associated Press. "Up to the middle of August," his message reported, "many of the canals of Mars discovered in 1877 by Prof. Schiaparelli were mapped here this year, but no one of them was seen to be double. On the night of August 17, Profs. Schaeberle, Campbell and Hussey made three entirely Independent drawings, each of which shows the canal marked 'Ganges' on Schiaparelli's map to be distinctly double." Thus, Holden concluded, "the Lick Observatory has the pleasure of confirming the discovery of Prof. Schiaparelli in 1892, as it already had done by its observation of 1890."[52]

Holden's telegram was a remarkable about-face from an astronomer who had previously complained loud and often about the unsuitability of newspapers for the announcement of astronomical findings. It was also a subtle massaging of the facts, phrased in such a way as to sound like the Lick had already, in 1890, confirmed Schiaparelli's double canals. In fact, Holden had until this telegram persistently maintained that though astronomers at the Lick saw Schiaparelli's canals, they had only rarely seen them double, and when they had such observations were never unanimous among his staff.[53] The shift to confirming Schiaparelli's doubling, or "gemination" as he called it, was a significant one. Not only did it thrust the canals, as a feature, even further into the limelight, but it did so by corroborating their apparent periodical duplication into parallel lines—a seemingly bizarre phenomenon that greatly enhanced the possibility that they might be artificial. Holden, of course, intended that his confirmation be taken as nothing more than an *observational* one, but, as might have been expected, the immediate press response was to speculate over what the double canals might *be*.[54]

This shifted the narrative of the ongoing Martian opposition in an important way. Whereas William Pickering's telegram from Arequipa had depicted Mars as a complicated, varied, and continually changing planet, Holden's telegram now focused attention on one apparently persistent feature. As the *Herald* wrote in response to the Lick's report: "The chief thing in the telescopic study of Mars now is an explanation of the 'canals' and of their gemination. It is hoped by many that while Mars is in a favourable position some observer will be able to clear up this mystery."[55] No doubt the *Herald* expected that observer to be William Pickering; sure enough, a week later the Arequipa astronomer obliged—of course unknowingly, being himself without newspaper access—sending a second telegram that, unlike the first, confirmed that "Many of Schiaparelli's canals have been seen single."[56] The fact that this simultaneously accorded with Holden's own claims (that Schiaparelli's canals were seen) and contradicted them (they were seen single, not double) only helped compound what was already a deeply ambiguous and confused discussion.

Yet for the press it was just this mystery that made the canals such a compelling and saleable news story. In the slow news days of the summer, papers across the United States and Europe readily seized upon the opportunity to spin out numerous articles and editorial notes about these strange dark lines. Much more so than any reports of melting snow or elaborate mountain ranges ever could, the conundrum of the canals more neatly fit the characteristics that newspapers looked for in a story: on the one hand, they were a single recognizable feature that could be returned to again and again, and, on the other, their very ambiguity in appearance and meaning allowed for a steady stream of exciting, varied, and speculative articles. In the weeks following William's second telegram, the newspapers buzzed with news of his "important data as to the curious straits or canals on [Mars's] surface."[57]

A conjunction of cycles—astronomical and news—created a sensation that did not settle down until the serious business of the coming presidential election took back the press agenda in the autumn. By then the Martian canals were already a cultural phenomenon, and one that even politics couldn't entirely escape. "If Mr. [Grover] Cleveland wishes to please the Republicans," wrote the *Herald* on September 15, "he will confine his letter to speculations concerning the canals of Mars and let American tin alone." The Democrats, another paper proclaimed, had "no more to do with [federal] money coming to [New York] State than they had with digging the canals of Mars."[58]

The bitter irony for Holden and William Pickering is that neither of them had intended their reports to be interpreted in this way. In their competitive pursuit of press exposure, they had both lost control of the narrative. Holden, ever the conservative observer, had always remained reticent to speculate in any way about the nature of Schiaparelli's dark, linear markings. Yet merely by confirming their existence he was drawn into very public debates he had no desire to be associated with. As he later noted in his end-of-opposition report for *Astronomy and Astro-physics*, "The present note may serve a useful purpose in correcting certain erroneous statements regarding our work which have been widely circulated and which require correction."[59] William Pickering was clearly much happier to speculate about Mars's geographic features and habitable conditions than Holden was, yet Pickering too had his doubts about the canals, having as early as 1888 suggested that they might actually be vegetation rather than waterways. His observations at Arequipa in 1892 only further convinced him that Schiaparelli's depiction of Mars was overly simplistic, and he wrote to his brother on July 7 that "I don't take must stock in Schiaparelli's canals on Mars. There are some coarse ones there undoubtedly, which were seen by others before him. There is also some extremely fine detail, which I never saw before, but I don't think it takes the form of canals, at least not as drawn by Schiaparelli."[60] William's observations, summarized in his telegrams to the *Herald*, were clearly intended to sweep away what he saw as a limited and flawed description and provide instead a completely fresh view of the planet, with the "canals" incorporated into a much more varied, complex, and dynamic representation. But by mentioning them at all, William played into a narrative that the newspapers had already begun to form, in which the canals were the central feature of an ambiguously living world.

The genre with which Holden and William Pickering engaged very much exacerbated this problem of interpretation. Newspapers were ideal because of their speed and their reach, but these features came at the expense of detail. Relying exclusively on the telegraphic network to move international astronomical news meant that the opposition of 1892 played out entirely without the exchange of drawing, photographs, or maps. As Holden had always stressed, it was the exchange and accumulation of these images of Mars that allowed astronomers to cautiously build up an accurate picture of what the planet looked like.[61] Working exclusively with words left a great deal more to interpretation, a problem further exacerbated by the impact that the telegraph itself had on language. The financial imperative for brevity produced a clipped telegraphic vernacular that was stripped of detail and that separated the observer from the writer. Required as both a transporter and a mediator, newspapers necessarily interpreted telegrams to make stories.[62] Detailed descriptions of intricate planetary detail and complex topographical changes could neither be conveyed using the telegraph nor were desired by newspapers looking for saleable copy. By the time William Pickering could publish his own drawings of Mars in a special issue of *Astronomy and Astro-physics* (figure 3.3), the great Mars boom was already winding down, and his visual representations of dramatic change over time did not reach a public consciousness that had already consumed the newspapers' depiction of Mars as a more or less static, canal-covered planet.

The ambiguities inherent to such newspaper interpretations were also amplified by the ambiguity of the word "canal" itself. What astronomers saw were dark, straight lines, at least twenty miles across and often thousands of miles long. As Schiaparelli himself noted, both brevity and conventional analogical terminology led him, for simplicity and clarity's sake, to dub these features "canali."[63] Ironically, William Pickering was among the first astronomers to note the possible danger of providing such puzzling features with such a concrete name, writing in 1890 that "it seems to me most unfortunate that the name of canals has been attached to these finer markings upon the planet, for there has not been the slightest evidence brought forward in support of the supposition that they are filled with water."[64] Even if they *were* water, as many astronomers believed, the name still provided a dangerous terrestrial analogy that was ripe for misinterpretation.[65] It is telling, therefore, that at the moment that this risk of misinterpretation became most pressing, thanks to the press furor over Holden and William

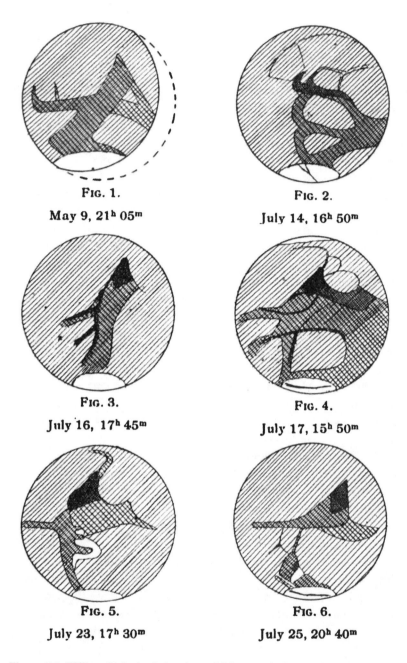

FIG. 1.

May 9, 21ʰ 05ᵐ

FIG. 2.

July 14, 16ʰ 50ᵐ

FIG. 3.

July 16, 17ʰ 45ᵐ

FIG. 4.

July 17, 15ʰ 50ᵐ

FIG. 5.

July 23, 17ʰ 30ᵐ

FIG. 6.

July 25, 20ʰ 40ᵐ

Figure 3.3. William Pickering's drawings of Mars, made during the planet's 1892 perihelic opposition and first published in his post-opposition report for *Astronomy and Astro-physics* (vol. 11, plate 30). Presented in series, Pickering's drawings show complex changes in the planet's physical appearance over the course of the opposition. Unlike his telegraphic reports, such images did not appear until well after the great Mars boom was already over and even then only reached a small, specialist audience.

Pickering's observations, the argument emerged that the appellation of "canal" derived from an unfortunate *mistranslation*.

Norman Lockyer, in an angry September 8 note in his own journal, posited that the translation of Schiaparelli's Italian might be the root cause of the sudden and unnecessary sensation. "These streaks he called *canali*, which in Italian, as *canalis* in Latin, means either a channel, a canal, or a pipe. Unfortunately, however, whenever it has been translated into English the word *canal* has been used, which of course with us suggests human labour. We have already seen what this has led to."[66] Mistranslation, therefore, which has subsequently been posited by almost every historical account of the "canal controversy" as its root cause, must rather be understood as a partisan explanation deployed by certain actors during the debate itself. The complex rendition of images as words via the telegraph and across mass media generated for some a crisis of translation, in which the meaning of observations risked being lost in the message. Mistranslation became one plausible way to explain away this problem of linguistic ambiguity.[67]

Newspapers, for their part, were happy to revel in these ambiguities. They need not worry if the word "canal" might infer artificiality, and they actively relished the terrestrial analogy that it implied. By taking what Holden and William Pickering had written and interpreting it for their own ends, they took a single feature, described by a single potentially problematic word, and placed in at the center of the 1892 opposition. Newspapers, more than any other medium, played an active role in transmitting the events of 1892 as they happened, shaping the public perception of Mars as a canal-covered planet. This canal narrative would be the enduring legacy of the great Mars boom.

EXPERTISE AND AUTHORSHIP IN EVENT ASTRONOMY

At the heart of event astronomy was a powerful alliance between internationally networked newspapers and new remote sites of astrophysical work. This novel media-labor dynamic was at once geographically specific and inherently serial: a few isolated, privileged sites presented competing claims, mediated through specific news outlets, in a concatenated progression of discrete events. A close analysis of debates over the physical constitution of Mars during the 1892 opposition suggests that two intertwined problems loomed large in this new astronomical order.

Astronomers, for their part, had to grapple with the question of how to present their observations in the public sphere. How should facts or discoveries made at a single specific site be rhetorically shaped into messages that could travel, near instantaneously, across the globe to audiences that potentially numbered in the millions and ranged from expert colleagues to readers of the penny press? Issues of broadcast speed, public exposure, disciplinary decorum, peer scrutiny, media format—even money and fame—were all at stake in finding an answer. When Edward Pickering wrote to his brother shortly before opposition to demand that he prioritize photographing stellar spectra over visual observation of Mars, William's reply cut right to the heart of the matter:

> I recently got a cable from the N.Y. Herald for 200 words on Mars. From this I presume there is some interest felt in the States on the subject. I had proposed reversing our hours shortly, to get on with the spectra, but it seems now as if it would be better not to do so at present, even if some spectra have to go over till next year. If I can find out enough, and arouse public interest enough, perhaps we can get the $300,000 for a 40 inch glass down here. If so, I presume that is the "most important investigation" I can undertake.[68]

For William the fame of discovery—of making the news—went hand in hand with the kind of publicity that put observatories on the map and made them money. But for Edward a fine balance between publicity and notoriety had been crossed, threatening his observatory's reputation. This was evidently a matter of judgment—not of *if* astronomers should nurture and exploit relationships with newspapers but of *how*, as Edward's reply to William makes clear: "[Your] telegram to the N. Y. Herald has given you a colossal newspaper reputation. A flood of cuttings have appeared, forty-nine coming this morning. In my own case I should have restricted myself more distinctly to the facts in this as in other cases. You would have rendered yourself less liable to criticism if you had stated that your interpretations were probable instead of implying that they were certain. A reexamination of your telegram shows that it is more guarded than some newspaper versions of it."[69]

Newspapers are clearly not innocent agents here. Edward's reply to William is a critique of both of the principal actors working beyond his control: his brother and the *New York Herald*. Disagreement over transmission practices was a two-way street, and the genre choices at the heart of the great Mars boom were choices for editors and journalists too. Newspapers—in all their immense diversity of readership, scope, politics, and style—themselves had to grapple with the question of how to present competing claims about Mars, faced with contested claims to authority and contrasting abilities to generate successful copy. This was in no small part because, whether they were stoking a sensation about extraterrestrial life or moralizing about the editor's role in safely mediating scientific information, they were—and saw themselves as—active participants in the astronomical work that they reported. The great Mars boom was, therefore, not only a sensation about Martian canals but also a sensation about scientific journalism.

Despite the obvious appeal of the canals, at least some journalists took time to worry about their new position as the discipline's principal gatekeepers of information exchange. Milicent Shinn, a personal friend of Edward Holden and a major player in California journalism, responded to criticisms of the Lick's initial Martian reports with an editorial broadside against her more sensation-mongering colleagues:

> The local newspaper criticism of the Lick Observatory in the matter
> of the observations of Mars during the opposition that has just taken
> place, seems to be extremely foolish, and some of it is written by ed-
> itors that ought to know better. Neither [myself] nor any other editor
> in California is competent to judge at first hand of the qualifications
> of an astronomer. . . . The reproach that the Observatory has not
> announced astonishing discoveries in Mars is something that rests
> on surface facts, and it seems to us childish. It may prove, when the
> returns are all in, that the Lick Observatory has done less with the
> opposition of Mars than other observatories, but nothing of the sort
> appears as yet. What does appear is that Professor Holden has sent
> out curt and cautious dispatches reporting the observations, where
> other astronomers have sent out long and interesting ones, containing
> speculation as to the causes of the appearances seen.

Their speed and reach placed newspapers squarely at the heart of event astronomy's ongoing conflicts, a position that raised serious issues of propriety and expertise, given that the medium's stated ambition was to sell copies, not to safely steward scientific information exchange. "Newspapers ought not lend themselves," Shinn concluded, "to such crude notions as that the use of great telescopes is to discover people and houses on the moon, canals and electric lights on Mars. . . . The editors that know this are the ones that should be entrusted with the editorials on scientific matters."[70] Astronomy and journalism's relationship was a symbiotic one, but it was also a problematic one.

I have argued here that a close analysis of this problematic relationship demonstrates the ways in which transmission technologies shaped the construction of new knowledge about Mars. This influence is evident in two complementary ways. First, within the tight time-frame of the immediate episode analyzed in this chapter—the boom of August and September 1892—we see how the linguistic limits, communication speed, and geographical scope of submarine telegraphy shaped and bounded what could be said about, and therefore what could be known about, Mars. The result was a particular focus on one persistent feature, the planet's canals, *despite* the intentions of William Pickering himself. This focus, therefore, was not merely a second-order effect of public reception. Because astronomers themselves, on all sides of the debate, came to rely so completely on newspapers such as the *Herald* and its Associated Press rivals to communicate their observations and arguments quickly and globally, telegraphic news became a constitutive part of astronomical work on Mars. Nowhere is this effect clearer or more striking than in the response of Edward Holden, who, despite decrying the rise of newspaper astronomy before the boom, responded to William's sensational reports with his own counter-report, telegraphed directly to the Associated Press, announcing the Lick's observations of doubled canals. The 1892 nexus of a very few remote giant telescopes, globally networked newspapers, and rival astronomers keen to secure status, priority, and reputation combined to construct the canals on Mars as *the* central feature of an ambiguously living planet.

Second, then, we must also attend to the long-term effects that this construction of a canal focus had on debates that continued well into the twentieth century. After 1892 astronomers of markedly differing views and skillsets configured their Martian work in terms of what the canals might be, while continuing to propagate this work in an extremely wide range of media types to a huge and diverse audience receptive to—as Agnes Clerke put it—"news from Mars." Thus it was that François Terby could celebrate, in the wake of the boom, the decisive defeat of "Schiaparellophobomania" among astronomers who had been on the fence about the existence of Schiaparelli's enigmatic features before 1892.[71]

No one, of course, tapped into this post-boom focus on the Martian canals with more alacrity or success than the man who would come to dominate the Martian canal controversy for the next two decades: Percival Lowell. Lowell was a late arrival to these debates, however, and his success must be understood in the light of the events of 1892. Indeed, the technological West and William Pickering's bold and spectacular work in Peru would form the template for Lowell's project, which began two years later. Working as a hybrid astronomer-author-adventurer on the Colorado Plateau, Lowell explicitly linked his self-funded "investigation into the condition of life in other worlds" to a focused study of one singular entity, "the canals of Mars," which he took at the outset to be "probably . . . the result of the work of some sort of intelligent beings."[72] Combating Percival Lowell would be the next great challenge for conservative professionalizing astrophysicists, and in this fight too mass media would prove crucial.

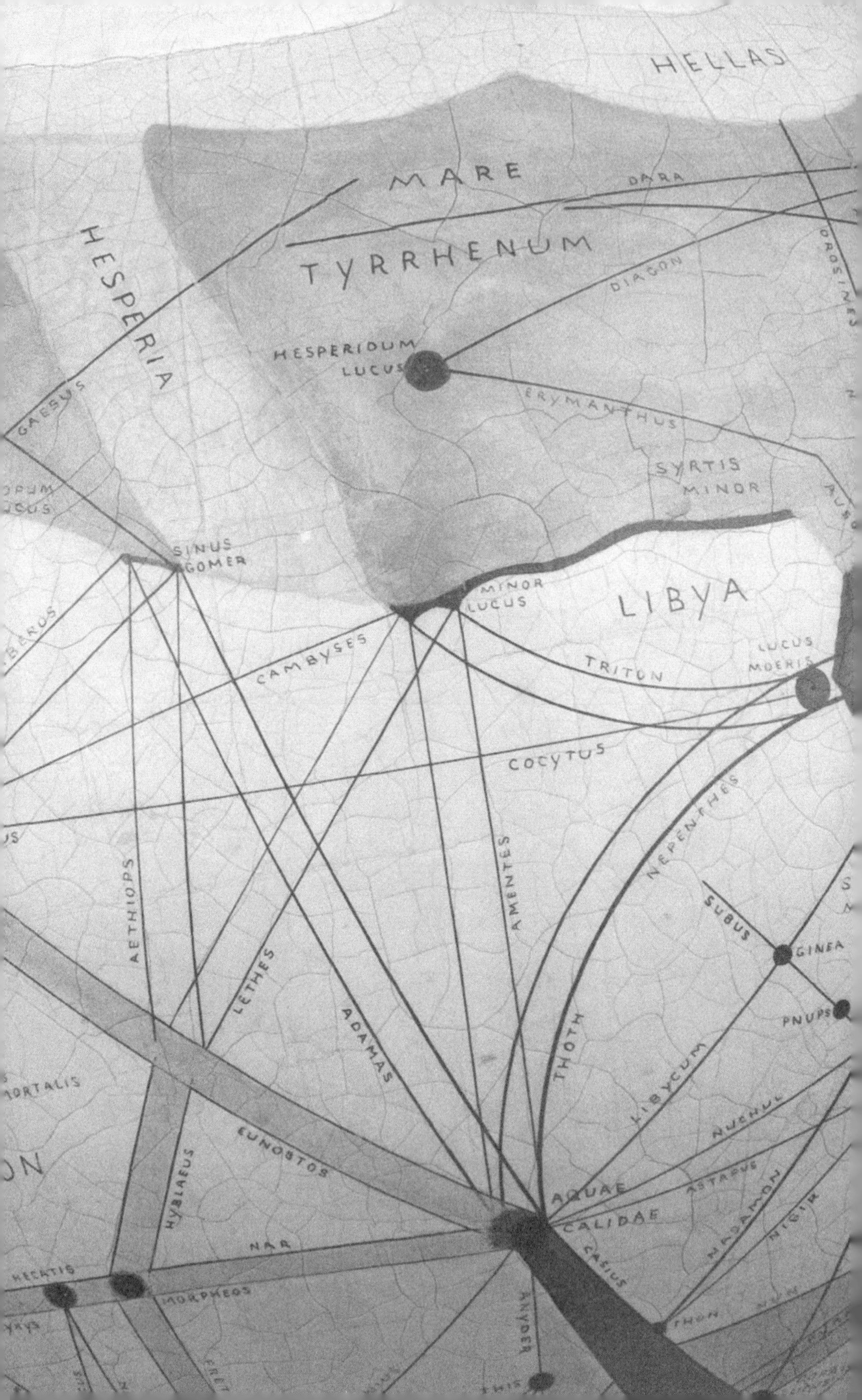

HELLAS
MARE
TYRRHENUM
HESPERIA
DIAGON
HESPERIDUM
LUCUS
ERYMANTHUS
SYRTIS
MINOR
GAESUS
SINUS
GOMER
MINOR
LUCUS
LIBYA
CAMBYSES
TRITON
LUCUS
MOERIS
COCYTUS
NEPENTHES
AMENTES
AETHIOPS
LETHES
ADAMAS
THOTH
SUBUS
GINEA
PNUPS
LIBYCUM
NUCHUL
EUNOSTOS
HYBLAEUS
AQUAE
CALIDAE
ASTAPUS
NASAMON
NIGIR
NAR
HECATIS
MORPHEOS
CASIUS
ANYDER

MADE TO LAST 4

"Mars" in the Eleventh-Edition
Encyclopaedia Britannica

In late February 1895, crowded audiences packed into Boston's Huntington Hall to hear Percival Lowell deliver four lectures on Mars. Just as Richard Proctor had a generation earlier, Lowell drew capacity crowds to hear "the very latest advices from that planet." Reporters on hand to capture the scene described Lowell's "charm . . . humor, his ready wit," and his "complete knowledge of the subject with which he deals." Lowell's lectures ranged across the founding of his huge new observatory near Flagstaff, Arizona, the observations made there during the 1894 opposition of Mars, and the "marvels" that this work had revealed. It was the astronomer's firm conviction, the *Boston Commonwealth* reported, "that intelligent beings occupy the planet Mars, who know how to work in the common good, who have contrived public works of vastly larger extent than we of the Earth have dreamed of, and have carried out their contrivances with a precision and strength wholly unknown in mundane affairs." Less than a year into his career as an observer, Lowell already had his local press convinced: "There are not more than twenty people in this earth who have seen what he has seen. Even some of the great observatories of the world are so situated that they have not noted the marvels which the Flagstaff observatory has revealed to us. But truth is truth, and it matters but little whether at this moment it have twenty apostles or two thousand. It is certain that the revelations which the Flagstaff observatory has made from its signal station to the world are revelations which will be accepted."[1]

Percival Lowell may have been late to the study of Mars, but he was not slow in tapping into a preexisting fascination with the planet and its canals engendered by fifteen years of reports, speculations, and events. Newspapers, particularly those with a proclivity for sensation, were already attuned to the popularity of fresh news from the red planet, and many eagerly printed the frequent reports that began to emanate from Lowell's new observatory. Lowell's great success, then, was by no means an unprecedented phenomenon. Considered in his proper chronological place, Lowell sits nearer to the end of a history of news from Mars than he does to its beginning. On these terms, it immediately becomes evident that one of his greatest skills was, in fact, in taking advantage of a preexisting news infrastructure, entering into and brilliantly exploiting an astronomical marketplace that had already been shaped to a large degree by the exploits of men like Richard Proctor and William Pickering. The template had been set, both by imaginative astronomy and in particular by the event astronomy sensation of the great Mars boom. By constructing a high-cost, high-altitude observatory, Lowell met the exacting requirements of the technological West, entering a very small club of astronomers who could claim Martian expertise by recourse to superior seeing. And as a successful, talented author and lecturer with a knack for promoting his novel theories, Lowell rapidly came to dominate the news economy of Martian astrophysics.

Lowell's success posed a serious problem for many of his colleagues, who soon realized that the problems of 1892 risked being not only repeated but exacerbated by Lowell's sensational new project. Not least, this was because William Pickering himself had been hired by Lowell to help establish the new Arizona observatory, which would be dedicated to the study of Mars. William, who had struggled to settle back in Boston under the immediate shadow of his older brother, Edward, jumped at the chance to return to the red planet, taking another Harvard Observatory employee, A. E. Douglass, out west with him.[2] News that Lowell had signed up as his sidekick none other than the impresario of the great Mars boom immediately provoked a wave of consternation from among the discipline's elite. Positional astronomer Seth Chandler wrote to his friend E. E. Barnard to lament that "Lowell, I am sorry to say, has not selected the right kind of companions for his astronomical picnic"; what he needed, rather, was someone "sound and conservative, to sit on his coat-tails and keep him down to business, and prevent wild flights of fancy." His choice of William Pickering all but ensured the exact opposite. "It is, unfortunately, perfectly safe to predict," wrote W. W. Campbell on the eve of the 1894 opposition, "that we shall also hear from the sensationalists, astronomers included; and that fact is a source of sincere regret to all healthy minds."[3]

These fears were soon confirmed in spades when Lowell launched the publicity tour for his new observatory in the spring. Speaking before he or any of his staff had made a single observation, Lowell confidently informed the Boston Scientific Society that "the most self-evident explanation from the markings [on Mars] is probably the true one; namely, that in them we are looking upon the result of the work of some sort of intelligent being."[4] Unsurprisingly, this claim immediately spread through the transatlantic press, angering Edward Holden enough to incite a furious editorial response in the Lick's house journal, *Publications of the Astronomical Society of the Pacific.* Not pulling any punches, Holden's riposte directly equated Lowell's recent public statements with what he saw as their most comparable and harmful predecessor: William Pickering's 1892 telegrams to the *Herald.* Just as in the case of Pickering's Arequipa dispatches, Holden warned, accounts of Lowell's lectures would circulate widely and be read by many "who are not sufficiently instructed in the details of [astronomical research] to form independent judgements." It was, he lectured, "the first duty of those writing for such a public to be extremely cautious not to mislead. . . . Conjectures should be carefully separated from acquired facts; and the merely possible should not be confused with the probable, still less with the absolutely certain."[5] The British journal *Observatory* soon added to the mounting consternation, endorsing Holden's critique and editorializing that "newspaper reports during the Mars *boom* of 1892 [stated] facts concerning the geography and climate of Mars . . . with a certainty about equal to that pertaining to similar facts about Central Africa." Six months later they went further, proposing an embargo on all newspaper astronomy: "To encourage the daily papers to look with any regularity for astronomical news of a popular character would seem rather dangerous to [astronomy's] best interests. . . . The tendency is, perhaps, already to publish results too soon instead of working them out thoroughly, and any encouragement of news-gathering would probably increase this tendency, and is therefore to be deprecated."[6]

Here was yet another crisis over the control of news from Mars. The interwoven problems of genre, disciplinary identity, speculation, geographical and instrumental priority, relations of astronomical centers to mediums of mass communication, and standards of appropriate rhetorical form in the public sphere were all once again copresent. At these problems' base remained one central question: who, precisely, had the authority to define the norms and bounds of planetary astronomy, and who, therefore, had the authority to tell Mars's public story? The success of Proctor's imaginative astronomy and the sensation of Pickering's Peruvian reports combined to make Lowell's own amalgamation of these paradigms, which he labeled his theory of "planetology," a persuasive answer for many. Lowell defined "planetology" as "the study of the planets of our solar system considered, not with regard to their gravitational action upon each other, but in respect to their own nature, constitution and history . . . as evolving worlds." The works of Proctor and Flammarion were formative influences here, and the kinds of shifts in perspective that they had pioneered played a central role in Lowell's theory. By eliding the study of Earth with the study of Mars, Lowell evoked an extraterrestrial perspective such that readers would "see ourselves as others see us" (see figure 4.1).[7]

As Robert Markley has shown, the success of Lowell's approach rested upon his great skill at tapping into multiple strands of contemporary popular scientific discourse, including cosmology, evolutionary biology, ecology, sociology, and the new technologies of communication and projection. By building his theory up from the two great secular theories of nineteenth-century science—the nebular hypothesis and Darwinian evolution—Lowell could pitch himself as a secular, progressive scientist using objective reasoning and analogy to challenge the dogmas of his conservative opponents. At the heart of this method was a compelling chain of reasoning that linked prepotent visual observations with a range of terrestrial analogies. Older and colder than Earth, Mars was a dying planet that, much like the impoverished colonial spaces of India and Africa, suffered from near perpetual drought and resource scarcity. But its inhabitants were also more evolved than their terrestrial counterparts, providing them with the intellectual faculties and technological prowess to engineer a brilliant, planet-encompassing solution: a network of irrigation canals capable of distributing polar meltwater to crop fields in Mars's warmer equatorial regions. Channeling both contemporary fears of ecological decay and middle-class reverence for techno-triumphalist infrastructure and political order, Lowell constructed a Martian landscape and society that was vivid and dynamic, could be imaginatively experienced, and could foretell the fate of Earth. Planetology was exciting, progressive, and persuasive, and it filled lecture halls and sold books and newspapers.

A

Figure 4.1(A–C). Percival Lowell's planetology deployed a range of visual techniques to establish a familiar geological and biological history for Mars. In his *Mars as the Abode of Life*, Lowell substantiated his narrative of Mars as an aged, arid, dying planet through analogy with terrestrial landscapes, such as image A, the Arizona desert local to his observatory (93). An alien perspective also helped link views down on Earth, as in B, Hyde Park photographed from a balloon (plate 4), with views of Mars, as in C, a planisphere entitled "Appearance of Mars in 1905" (plate 7).

Importantly, planetology also complied with what most onlookers took to be the bounds of acceptable scientific practice.[8]

Proving Lowell wrong therefore meant dismantling a strong and deeply embedded popular perception of Earth's and Mars's places in the cosmological order. It also meant tackling a radical asymmetry in publicity. Lowell's ideas made news, and his trope of a dying Mars was soon incorporated into a huge range of popular commentary on the planet, not least the rapidly flourishing genre of scientific romance pioneered by H. G. Wells.[9] Lowell himself, meanwhile, commanded growing respect as a convincing and powerful spokesperson for his science, comfortably working across an exceptionally diverse range of media that spanned from the *Proceedings of the Royal Society* to *Popular Astronomy* to the *Atlantic Monthly* to the numerous newspapers

B

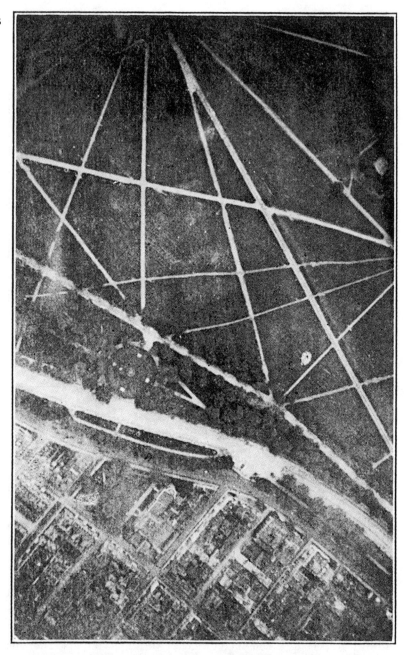

and periodicals that assiduously covered his astronomical investigations. As with Proctor, it is important to recognize all of this work as a form of astronomical practice rather than mere "popularization," and it should therefore not be overlooked that Lowell was active among

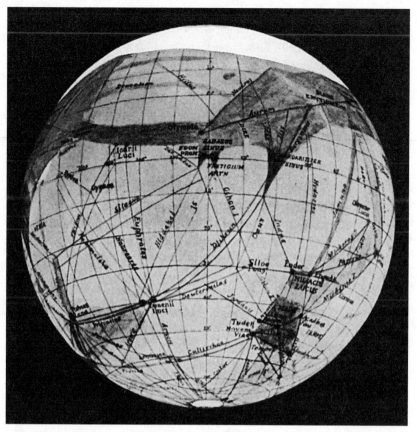

his disciplinary peers too.[10] The "Boston Brahmin" traveled widely to present and debate his ideas with the astronomical societies of Europe and the United States, and he ultimately secured a post as nonresident professor of astronomy at MIT, subsequently insisting upon the title of professor.[11]

This chapter is about how Lowell's opponents fashioned a plausible response to this powerful and diverse methodology. Lowell's own activities have been very well covered in the secondary literature, but, precisely because his persona has dominated the Mars canal controversy so completely, little focus has been paid to his rivals on the other side of the debate.[12] Analysis of transatlantic astronomy in mass media reveals one significant characteristic of this response. Because Lowell's opponents were disadvantaged by the inherent asymmetry of news journalism's coverage, one viable tactic became the employment of forms of mass media markedly different to those commanded by Lowell. As a response to the *immediacy* of periodical and newspaper coverage of

Lowell's planetology, opponents focused their ripostes through forms of media that were valued above all for their *permanence*. Because they were made to last and because they commanded massive audiences, it would be encyclopedias above all that critics of planetology would turn to. These opponents were therefore no more distant from mass media than Lowell himself. Rather, they were close to a different kind of media. A familiar skeptic would prove to be the most important figure in this response.

NEWCOMB IN THE NEW AGE

Simon Newcomb was sick of hearing the question. "What," he would be asked, "do you think of the canals of Mars?" The best answer he could give, he informed his readers, is that he "did not think anything about them." By "canal," he knew, the inquirer really meant "an artificial waterway," but "no astronomer ever saw or thought he saw anything of the kind." In point of fact, the widespread and unfortunate notion that anything like canals had been observed on Mars sprung from nothing more than a translation error. "If Schiaparelli had used any other modern language" than Italian, Newcomb lamented, "a great popular delusion" would have been avoided. It had not, however, been avoided, and as a result Newcomb now found himself in the front ranks of a firmly entrenched defensive force, squared off against an opposition marshaled by one powerful commander. As he drily put it in his *Harper's Weekly* article "Fallacies about Mars," "there is indeed a great difference of opinion between Professor Lowell and those who share his views on the one side, and the rest of the scientific world on the other, as to what is going on upon Mars."[13] Although this might look like an effective blow landed against his popular opponent, at the moment that Newcomb wrote, in the summer of 1908, it likely felt to him and his allies more like a rare counterstrike from a dwindling and ill-equipped company. Lowell was in the ascendancy, and Newcomb found himself at the forefront of a challenging resistance.

At first blush it is not surprising to find Richard Proctor's most stubborn and tenacious American sparring partner at the head of the queue to take on Lowell. Simon Newcomb was at the peak of his powers by the time of Lowell's ascendancy, a dominant figure in both transatlantic astronomy and rational public discourse. Happy to write for a wide range of popular periodicals and rarely content to stay out of matters of broad public concern, his stern and exacting judgment could be found cast over an exceptionally wide range of subjects, including childhood education, economic theory, the swindles of life insurance, the failures of psychical research, the relationship of theology to science, and the impossibility of powered flight. In all of these debates and many others besides, Newcomb championed the primacy of scientific expertise and the efficacy of the scientific method as palliatives for a broad range of social, political, and economic ills.[14]

Yet, by the early 1900s, Newcomb was also an aging positional astronomer in the age of the tyro astrophysicist, and his pivotal role in the Martian life debates therefore deserves closer scrutiny. A retired mathematical savant, Newcomb might actually be considered a somewhat unlikely foe for Lowell. His lack of interest in—indeed often outright disdain for—planetary astronomy certainly makes his entry into the canal debates far from self-evident. In what follows, I argue that it is his activities as a public intellectual that explain this apparent anomaly. Newcomb's broad concerns over astronomical authority, disciplinary control, and the mediation of knowledge claims to a mass public all proved pivotal in his response to Lowell's planetology.

As his clashes with Proctor make clear, Newcomb was profoundly intellectually conservative, even by the standards of mathematical astronomy. His austere mathematical mind was legendary, and he became something of a cultural touchstone of the fin de siècle—his name a byword for mathematical rigor and logical rationality. He was "too sound a mathematician" to be of any help in Henry Adams's pursuit of education, but he was suitably recondite to be the foil for the four-dimensional geometry of H. G. Wells's Time Traveller.[15] Yet by 1900, even nonspecialists like Adams could see that the old positive certainties of the physical sciences, so cherished by Newcomb, were beginning to wash away under a wave of mysterious forces, manifest above all by the electrical, spectral, and x-ray phenomena then much on display at the era's grand public exhibitions (see figure 4.2). Looking back on his experience of Paris's spectacular Universal Exposition, Adams could only conclude that "man had translated himself into a new universe which had no common scale of measurement with the old."[16] This common scale was the bedrock of Newcomb's science, and he found these transformations deeply troubling as a result. A firm admirer of Auguste Comte and a committed positivist, he had always insisted that scientific claims must be rooted in definite sensory experience. When excitement over radiation was at its peak, he stoically wrote to Lord Kelvin to express "a general lack of full confidence that the explanation of these obscure ethereal and molecular phenomena

Figure 4.2. The Palace of Electricity, Paris Universal Exposition of 1900, as illustrated in Boyd, *The Paris Exposition of 1900* (235). Forty-eight million visitors to the Paris exposition gazed upon a tumultuous world of scientifically revelatory machines and devices, including giant dynamos, automobiles, radio transmitters, x-ray machines, airships, chromophotographs, moving pictures with sound, and electric trams.

are within the grasp of our intellects. We can express the agencies in the language of algebra so as to work out their effects, but I am not sure that the agents themselves may not lie outside the limited range of our faculties, which know the outside world only as it is revealed through our limited organs of our sense."[17] These principles would ultimately become formalized into Newcomb's "creed of naturalism," a universal methodology that insisted on explicit evidence only and the deployment of facts expressed in clear and exact language. "In its relations to progress," Newcomb wrote, "the first work of science is to eliminate from our mental activities everything that has not the attribute which may be called practical. . . . It rejects all vague speculations, all meaningless forms of language; it uses no words which do not admit of exact definition in terms of actual experience."[18]

Newcomb accordingly took great pains to try to police both the speculative excesses of many of his peers and what he saw as the wayward irrationalism of wider society as a whole. His advocacy for a mooted "national research university," for example, envisaged the new institution as nothing less than a research engine for the "reducing [of] the laws governing the action of causes upon the social organization to a science which shall be exact in thought and method." For Newcomb there was one scientific method, applicable in principle to fields outside natural science and capable of fostering "right thinking" in a vast range of subjects, up to and including social organization itself.[19] In his turgid, unsuccessful foray into writing fiction, *His Wisdom the Defender* (1900), Newcomb went so far as to construct his own fantasy of what this model of progress might look like, set in an ordered scientific utopia of 1941 in which "two generations had passed without a scientific discovery that could be called epoch-making." His story's heroes are dutiful Anglo-Saxon college athletes, ruled by a righteous and rational-minded physics professor, who together harness the power of "etherine" to counteract gravity and build airships, from which they dispense republican rule and positivist empiricism across the globe. The enemy of this virtuous moral order is a sneaky, red-headed reporter for—who else?—the *New York Herald*, who threatens this project of enlightenment by peddling half-truths and rumors in the base pursuit of a popular audience.[20] Such a fantasy clearly epitomized Newcomb's own worldview, linking as it did scientific exactitude with a Teddy-Rooseveltian ideal of manliness and honor, and contrasting these with the imprecise and corruptible world of common knowledge. Much of his retirement was spent acting out a somewhat more mundane version of this vision; rather than college athletes in flying machines, Newcomb's weapons of choice were books, journals, public lectures, and above all the parallel enterprises of grand expositions and encyclopedias. Newcomb would become a key organizer of the scientific content for the 1904 Saint Louis World's Fair, and over the course of his career he would author at least two hundred articles across nine different encyclopedias, serving as an associate editor for five of them.[21]

Encyclopedias and expositions made sense as a platform for Newcomb's positivist project precisely because they were, in the right hands, powerful instantiations of ordered public display. As Rebecca Solnit has noted, the universal expositions that were at their fashionable height in the fin de siècle were, in a certain sense, "encyclopedias come to life."[22] Both enterprises aspired to a panoramic vision of modernity, and both shared a belief that the entire world could be collected, catalogued, and displayed. At the root of each was a fusion of popular education, commercialism, and monumental scale that exemplified public enterprises of the machine age and that fit the public prerogative of astronomy very well.[23] Crucially, the extravagant scale and scope of expositions and encyclopedias positioned both as viable competitors, in terms of audience, to the era's only comparable medium of mass communication, the newspaper. Expositions were typically visited by millions—sometimes tens of millions—of people, commensurate in scale with major city newspapers and encyclopedias, both of which could count their readerships at least in the hundreds of thousands.[24]

Distinct to the exposition and the encyclopedia, however, was their producers' aspiration that they be all-encompassing surveys of human knowledge. Rather than trading in ephemeral news or topical information, the exposition and the encyclopedia aimed at systematic reflection upon the state of human art and science. Their ideal form would present every aspect of human existence in a single carefully classified arrangement. They were, therefore, one key response to crises of representation induced by the nineteenth-century proliferation of fragmentary knowledge and information in the era's rapidly expanding tangle of books, journals, and newspapers. The bold ambition of the encyclopedia and the exposition offered the possibility of establishing definite, durable, permanent knowledge, a characteristic recognized and exalted by those who organized and chronicled them. As one exhibition organizer commented in 1902, the ideal exposition would be nothing less than an "encyclopaedia of the civilized world." Encyclopedias, in turn, captured in print the same cornucopia of scientific modernity that drew millions to expositions in a format that could be taken home, perused, and displayed as part of the emerging middlebrow home.[25] Not surprisingly, both these mediums of knowledge rationalization and display held great appeal for Newcomb, a man who decried the disordered cacophony of public science in the United States and yearned for the establishment of elitist, ordered control over its conduct.

The significance that scientists in this era placed on the content of expositions and the production of articles for encyclopedias should not, therefore, be underestimated. Richard Proctor, a prolific author of encyclopedia articles—including the "Astronomy" article for the ninth-edition *Encyclopaedia Britannica* (1875–1889)—closed his final diatribe against Edward Holden with the telling accusation that "it is well known my real offense with Prof. Holden was that I wrote the astronomy of the *American Cyclopedia*."[26] It was more than merely cachet that made such work matter. The massive audience for encyclopedic works and their recognized utility as tools for self-education placed them at the forefront of astronomers' interactions with their publics.

Holden's apparent anger over the *American Cyclopedia* probably had as much to do with Proctor's imaginative proclivities as it did with money or credit, and this same concern over content control and authority no doubt drove Newcomb's extensive encyclopedic labors.

The authority that practitioners of science hoped to establish for themselves and their synoptic accounts was symbiotic with the authority encyclopedia publishers and editors hoped to establish for their publications by employing preeminent experts as authors.[27] Simon Newcomb was such a draw in this regard that many of the myriad pirate, knockoff, and cheap imitation encyclopedias of the era, such as the *Twentieth Century Encyclopaedia* (1901), scurrilously listed the astronomer as an "associate" or "advisory" editor, much to his chagrin. The ploy worked, of course, only because Newcomb *did* write for numerous encyclopedias and dictionaries, which readers were expected to pick out as genuine articles among a sea of counterfeits.[28]

Recent work at the junction of book history and science studies has demonstrated the importance of such "routine authorship" in the making of knowledge and the shaping and maintenance of disciplines in the sciences.[29] Richard Yeo's seminal work on early Victorian encyclopedias in particular points to the important nexus of experts, editors, and disciplines that emerged around encyclopedic work in the first half of the nineteenth century. Yeo links major shifts in the format and rhetoric of encyclopedias such as the *Britannica*, especially their abandonment of any pretentions toward a grand scheme of classification in favor of alphabetical order, to the contemporaneous transformation of the sciences from natural philosophy to a collection of modern scientific disciplines. Central to such transformations was the employment of expert authors by encyclopedias for the production of large, singular treatises on specific scientific subjects, a task that promoted the definition of the intellectual bases of particular disciplines and their boundaries. Disciplines were "presented as a whole to be read as such." Yeo's analysis ends at midcentury, however, and therefore like most other encyclopedia histories it pays little attention to subsequent developments within the genre that were, I suggest, just as significant for the issues of expertise, authorship, and disciplinarity, at least with regard to astronomy in the late nineteenth and early twentieth centuries.[30]

My focus here is the eleventh edition of the *Encyclopaedia Britannica* (1910–1911), for which Newcomb served as the associate editor for all astronomy content. Entirely fittingly, the *Britannica* at this stage in its long life was the crowning embodiment of transatlantic print culture. Owned by a pair of New England press entrepreneurs, Horace Hooper and William Jackson, the eleventh edition was managed from New York, edited from the *Times*'s office in London with assistance from a New York suboffice, published by Cambridge University Press, typeset in the *Britannica*'s birthplace of Edinburgh, and printed in five locations across England and Scotland (and, for the American market, in Chicago), with the finished work dedicated jointly to both the reigning sovereign of Great Britain and the president of the United States.[31] As W. T. Stead would have predicted and delighted in, there was a recognized "Americanization" of the work in comparison to its all-British predecessors, of which Newcomb's involvement might be taken as just one example. Yet the choice of Newcomb, as the *Britannica*'s editor in chief Hugh Chisholm explained, was driven above all by the publication's desire to employ "our greatest living astronomer," and a similar ambition across the arts and sciences secured a transatlantic who's who of preeminent authors. "I'm determined that the eleventh edition must be the greatest book ever published," Hooper declared, an ambition that would set him back at least $1.4 million in production costs alone.[32]

The specifics of this work's novel features and overall structure—its internationalism, its grand scale, its comprehensive scope, its fragmentary architecture—all drew from and in part mirrored the expositionary culture of the fin de siècle. By 1900 the recognized value of exhibitions as a forum of public education had also raised pressing questions about their role and potential in representing science to a mass public audience.[33] As commercialism and spectacle grew ever more prominent, the educational prerogative and aims of such fairs shifted, especially as displays diversified and less centrally organized exhibitions were dispersed across multiple sites. Much like the reading experience provided by the many new highly illustrated mixed-topic periodicals of this era, visitors to expositions were presented with and soon became accustomed to a cornucopia of intermingled visual and textual representations, which they were expected to learn from through a process of visual "grazing." "In the new age" of 1900, historians have noted, "the public were treated less like inquisitive students and more like amazed spectators."[34]

A significant but overlooked transformation in the *Encyclopaedia Britannica* echoed these changes. For the eleventh edition, compilation of which began in 1903, the encyclopedia abolished the single discipline-spanning treatises for its science subjects, replacing them with dozens, sometimes hundreds, of shorter subject-specific articles.

For the first time the entire encyclopedia was also to be compiled at once and published in one complete panoptic issue.[35] These changes had a major impact on the temporal and disciplinary makeup of all scientific content. The previous (ninth) edition of the *Britannica* had explicitly repudiated involvement in "active controversies of the time," and, as Yeo has shown, it was a *historical* framework that underpinned the discipline-defining impact of these earlier editions' scientific treatises.[36] In contrast, the eleventh edition reflected turn-of-the-century public expositions by approaching its subjects as numerous discrete objects of cutting-edge analysis, to be displayed in a whole that readers were invited to explore at their own leisure. As one review put it, "all historical narrative, hitherto broken off at a safe distance from the present, is now brought down to the very eve of the year of grace in which we live."[37] In preparing the astronomy content for this new edition, Newcomb declared Proctor's ninth-edition treatise on "Astronomy" to be "antiquated" and therefore unsuitable even as a starting point for his own attempt to convey the subject, which would now come through more than 160 separate articles and definitions, including for such modish subjects as "Spectroheliograph" and "Chromosphere."[38]

This new format and structure greatly altered the nature and importance of the role of advisory editor. Rather than compose a synoptic historical account of his field, Newcomb's role for the eleventh edition centered on managing the often-contentious process of commissioning these articles. As Chisholm's preface to the new edition made clear, the principal reason for breaking down scientific content was the tendency for each "separate scientific question" within a discipline to now have "a single specialist" with "unique authority" in that domain. In the face of specialization, the single treatise was not only "cumbrous"; it was also unfeasible, tending as it did to omit certain "specific issues," particularly those at the leading edge of the science.[39] As a result, the management of articles on astronomical topics that were still under debate became a much more significant task for Newcomb than it had for his predecessor, Proctor. With appropriate experts to be commissioned and others to be placated, Newcomb's principal concern therefore shifted away from boundary work in competition with other disciples, toward an *internal* process of disciplinary control through editorial oversight. Such a task, offering as it did a great deal of power over his colleagues, was naturally relished by the imperious Newcomb. But it also presented significant challenges, never more so than in the case of the article on Mars.

ASTRONOMY AND VISUAL REPRESENTATION IN THE EARLY TWENTIETH CENTURY

Newcomb's first major job for the *Britannica* was as editor for the astronomy content of the tenth-edition supplements (1902–1903). Hastily compiled, these were eleven volumes added to the end of a reissued ninth edition, revising those treatises deemed in need of an update. This work served as a proving ground for Newcomb's much more substantial work on the eleventh edition, and its most notable consequences were twofold. On the one hand, links were forged between Newcomb and a small team of mostly American authors (including Edward Holden and George Ellery Hale) that he deemed trustworthy contributors on astrophysical subjects. On the other hand, the work inaugurated Newcomb's conflict with Lowell over Mars. This conflict would come to hinge on the contested status of two orders of visual representation: that of the astronomer themselves, as they drew what they saw at the telescope; and that of public exposition, in which the tropes of imaginative astronomy loomed large.

In the tenth edition's revised "Astronomy" treatise, Newcomb wrote dismissively of the evidence for canals on Mars, repeating Lockyer's mistranslation claim and suggesting that there was "by no means a complete agreement" between different observers' accounts. Newcomb gave the final word to the Lick's E. E. Barnard, whose generally overlooked *Monthly Notices of the Royal Astronomical Society* report citing disparate detail observed on Mars during the 1894 opposition was given the majority of space, a full citation, and the endorsement that Barnard was both "conscientious" as an observer and technologically advantaged by the Lick Observatory's "great aperture." Newcomb himself was inclined from the outset to implicate flaws inherent to the "psychology of vision" as the root cause of the discrepancies between Barnard's "minute, intricate, and abundant" markings and Lowell's linear dark lines. Citing an article on visual perception by the Italian astronomer Vincenzo Cerulli, Newcomb suggested that it was "quite natural that such features as those described by Barnard should, when seen with less sharpness, present to the eye the appearance shown on the maps of Schiaparelli."[40] In his popular book of the same year, *Astronomy for Everybody*, Newcomb explained this problem of perceptual conflation in greater detail. If one imagined examining with a magnifying glass a "stippled portrait engraved on steel," he wrote, "nothing will be seen but dots, arranged in various lines and curves." But remove the magnifier and "the eye connects these dots into a well-defined collection of features representing the outlines of the human face." Just as "the eye makes an assemblage of dots into a face, so may it make the minute markings on the planet Mars into the form of long, unbroken channels."[41]

Upon reading Newcomb's piece for the *Britannica* tenth edition, Lowell immediately wrote two long letters of complaint to his colleague, presenting to him "some facts which *prolonged observation* has taught me with regard to the markings on Mars." Lowell gave Newcomb a long list of arguments, most of which would soon become ubiquitous in his many defenses of his observatory's output. These were (1) that Lowell and his staff had the most experience observing Mars; (2) that only certain observers had the right sort of "acute" eyesight for the perception of fine detail; (3) that the mass of detail seen by Barnard was a symptom of the Lick's poor atmosphere, which inhibited seeing conditions; (4) that under "the very best of seeing" conditions at Flagstaff, such detail was resolved into distinct lines and dots; (5) that Cerulli's criticisms were outdated; (6) that experiments on the visibility of telegraph wires had shown that lines as narrow as two and a half miles should be perceptible on Mars; (7) that the reports of double canals as "diffuse streaks" by "many observatories" was due, again, to "poor air"; (8) that "negative testimony is no evidence at all"; and (9) that diversity among canal observations was positive evidence of the planet's complex life cycle. Newcomb's reply was diffident, conceding that though they differed "in the matter of interpretation of detail," he had "not had experience enough to criticize what you say about the seeing of the mass of details, but it looks . . . possible."[42]

Newcomb's attitude at this stage in the debate seems somewhat ambiguous. Though he was skeptical of Lowell's extravagant claims, he was also a firm believer in the priority of scientific experience, and as an outsider to debates about the physical constitution of Mars he was inclined at first to be relatively open to the workers at Flagstaff. By the early 1900s a small coterie of astronomers had begun to form an alliance against Lowell and his canal claims, most notably W. W. Campbell and E. E. Barnard at the Lick Observatory and George Ellery Hale and Edwin Frost at the Yerkes Observatory.[43] Newcomb at this time was not nearly as critical as these four, as evidenced by his inviting Lowell to take part in the first conversazione of the newly founded Astronomical and Astrophysical Society of America. Hale scolded Newcomb for not consulting with other members of the society's council before extending the offer, warning him that such "indirect indorsement" would only "create a demand for the sensational, and leave the public dissatisfied with the legitimate output of the large observatories."[44] Newcomb, in contrast, though he stood by his brief *Britannica* supplement, moved to reassure Lowell that his own lack of expertise in the subject—especially in comparison with the seasoned Martian observers at Flagstaff—made him far from absolute in his doubts:

> I highly appreciate the pains you have taken to set me right in the matter of Mars. As it is uncertain whether I shall ever have anything more to say in print on the subject, it is hardly worth while to discuss my views at length, but you seem to misapprehend the exact nature of my conclusions, if such they can be called. My general principle has been to take a sort of average of the general results of observations and of the different views as to what can be seen on the surface of the planet. Doubtless if I had to repeat them, I would modify some of my expressions in the new light thrown on the subject by your letters. It seems to me that, if there is a serious difference between us, it is only of interpretation. I have never claimed that the channels on Mars are mere illusions, but only follow Cerulli in the belief that, if we could observe Mars at 100th its present distance, we should put a different interpretation on the appearance presented.[45]

Newcomb's critique, as tentative as it was, was clearly solidifying around the issues of visual perception and the danger of perceptual conflation. That there might be problems with observers' interpretations of what their eyes perceived at the telescope was by no means a new objection. Proctor had written for *English Mechanic* as early as 1872 on "Self-Deception in Observation," noting the potential influence of preconceived opinions on observed results.[46] Others, meanwhile, began to ponder the simple weakness inherent to human vision, noting that detail below a certain level tended to be conflated by the mind's eye into features and shapes that were not present upon closer inspection. The earnest Greenwich astronomer E. Walter Maunder, in particular, was a committed proponent of this problem with respect to Martian observations. As early as 1894, he had noted that in the case of the canals, "we cannot assume that what we are able to discern is really the ultimate structure of the body we are examining."[47]

Lowell was highly attuned to this issue from the outset of his work, and he labored hard to forestall criticisms of his and his staff's perceptive faculties, on several grounds. Privately, this included his initial encouragement of Lowell Observatory assistant A. E. Douglass in the undertaking of a range of experiments using "artificial planets" to probe the optical effects caused by instrumental and atmospheric shortcomings. Lowell soon halted these tests, however, when, according to Douglass's bitter recollection, "it was evident [they] cast doubt on some observatory publications." (Douglass soon grew disillusioned with Lowell's dogmatism and was summarily fired in 1901 after he had written to Lowell's brother-in-law urging him to try to persuade Lowell to reduce his speculative tendencies.)[48] Publicly, Lowell echoed Hermann von Helmholtz's celebrated investigations of the role of experience and training in perception, arguing forcefully that "the expert sees what the tyro misses, not from better eyesight but from better mechanism in the higher centres." Simultaneously, and somewhat contradictorily, Lowell also publicized an idea he had shared privately in his letters to Newcomb, that there were broadly two types of astronomical eye, "sensitive and acute," with only the latter type (shared by himself and Schiaparelli) suitable for the minute definition and precise perception of objects essential for planetary observation.[49]

Maunder would be the first astronomer to mobilize against these claims. A deeply pious observer who had grown exacerbated by Lowell's bold pluralist assertions, Maunder teamed up with a Greenwich schoolmaster in early 1903 to conduct a rigorous set of visual perception experiments specifically designed to ascertain "the actuality of the 'Canals' observed on Mars." This now-famous study, conducted on schoolboys, demonstrated to the experimenters' satisfaction the strong tendency of observers beyond a certain distance to draw coherent lines, even when none were actually present on the image of Mars being viewed.[50] In the heroic myth of Lowell's defeat, this study features prominently.[51] Its impact at the time is much less clear-cut, however. Camille Flammarion soon repeated the Greenwich experiments on Parisian schoolboys and got the opposite result.[52] Lowell also mobilized a battery of tests to demonstrate contradictory results, then went on a publicity blitz to highlight the logical weakness of Maunder's claims. By using "blank slate" observers (untrained schoolboys), Maunder had hoped to produce objective results, but as Lowell made abundantly clear, the subjectivity inherent in an astronomer's drawings was, in fact, precisely the point in question. This question of skill was an understandable point of pride among astronomers. As Asaph Hall put it to a journalist, "People who are unaccustomed to looking through a telescope can rarely discover things which to astronomers are very distinct and easy to locate." So the challenge was actually to discern where this skill ended and artifacts of vision began.[53] This was not something schoolboys appeared able to answer. In a letter of response to Maunder read before the British Astronomical Association, Lowell asserted that because "*a* may produce the effect of *b* furnishes no proof or even presumption that *a* is *b* since *b* itself would produce the same effect. . . . This is a matter of astronomical experiment and not of psychologic discussion." Lowell made the vanishing point of distant lines viewed through telescopes the relevant calibration for the trained eye, rendering psychological experiments based on disparate detail irrelevant.[54]

Although Newcomb is reported to have "strongly supported" Maunder's 1903 paper, at the time he did not comment further, either to Lowell or in public.[55] In fact, as he made clear in his March 1903 letter to his Flagstaff colleague, he initially had no intention of further debating or writing about Mars and its canals. But Newcomb's wider work in this period kept the key issues here—of astronomy, expertise, and vision in public discourse—very much front and center in his life as a public intellectual. The two years following the tenth-edition *Britannica* work would be particularly busy for him, encyclopedic visions again proving an irresistible lure. In 1903 he traveled to Europe to drum up support and attendees for the Louisiana Purchase Exposition, to be held in Saint Louis in the summer of 1904. Newcomb was president of the Congress of Art and Science attached to the exposition, placing him in charge of one of the era's largest public gatherings of scientific expertise. Classified, like the latest encyclopedias, according to an overarching synchronic arrangement, the exposition was lauded by one commentator for its "deep, far reaching, ethical, and educative import."[56] At the same time, Newcomb also worked on these projects' print equivalents, completing his *Astronomy for Everybody* in 1902 and acting as an editorial advisor and author for the revised *Encyclopedia Americana* (1902–1906), the latest New World rival to the *Britannica*. These various encyclopedic visions were often explicitly linked, with the *Americana*, for example, carrying reports from the section directors of the Saint Louis Congress, so as to "preserve in permanent form the results of the Exposition."[57]

Across all these projects, Newcomb remained vigilant in defending the strict application of his "creed of naturalism." Writing to accept the *Americana* job, he made clear to its managing editor his view of what a proper encyclopedia should be: "a place where one can look readily for condensed statements of *facts* and *principles* on every subject of human inquiry." When the editor pondered the removal of tabulated astronomical data in order to save space, Newcomb exploded with indignation, retorting that "if the Encyclopedia is not intended, as I supposed, to include as complete statements of fact as possible . . . I do not see for whom it can be intended, except for the unintelligent classes, who will never want such a book anyhow. . . . Of course your authors may not like this, because it is so much easier to write a Sunday morning newspaper article than it is to give exact information which requires reference to books."[58]

Inherent to Newcomb's model, therefore, was a fundamental repudiation of all forms of scientific information that could not stand up to positive empirical test. This made the Sunday morning newspapers a lost cause—a genre without hope of establishing or promoting knowledge of a positive kind. Hence the vital importance of policing encyclopedia and expositionary content. For here too, as Newcomb knew all too well, rival, predominantly *visual* epistemologies of astronomy vied for popular attention. Imaginative astronomy in particular, with its prioritization of drama and excitement over hard empirical facts, was much on display in this era, from Paris's celebrated projecting "Great Telescope" to Georges Méliès's fantastical films to the kinaesthetic experiences offered by wildly popular spaceflight simulation shows (see figure 4.3).[59] Newcomb's vision for the Saint Louis congress and for the many encyclopedias he stewarded was as platforms devoid of such content. As a response to such visual spectacle, he offered instead a positivist scientific methodology predicated on the unambiguity of published facts.

FINDING AN AUTHOR FOR "MARS"

Mars's canals made trouble for this project. Within Newcomb's epistemology, the existence of life on other worlds was simply unknowable, humankind's knowledge of the universe being "infinitesimal when compared with the range it will have to include before anything *positive* can be said on the subject."[60] Yet the canals themselves, as real entities observed by multiple people, were, according to Lowell at least, empirical evidence indicative of a planetary lifecycle. Encyclopedia work would press the issue. In early 1905 work began on the astronomy content for the *Britannica*'s monumental eleventh edition, with Chisholm allocating to Newcomb 150 pages and a budget of $2,250 for completely revised material.[61] For the first time Mars would require a separate article, presenting Newcomb with the thorny task of securing

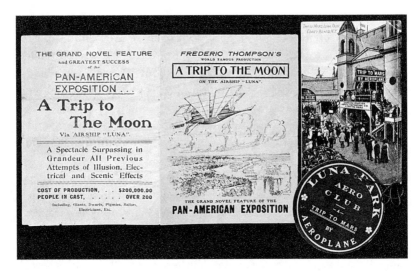

Figure 4.3. Ephemera relating to Frederic Thompson's spaceflight simulation shows "A Trip to the Moon" and "A Trip to Mars by Aeroplane." Thompson's moon ride debuted at the 1901 Pan-American Exposition in Buffalo, New York. Trips were made by a combination of electrical, scenic, and projection special effects, producing the sensation of flying through space and landing on the moon. In 1903 Thompson's ride formed the centerpiece of his new Luna Park amusement park on Coney Island. In 1910, following renewed sensation over the question of life on Mars engendered by the 1909 perihelic opposition, the ride was rebranded as "A Trip to Mars by Aeroplane," helping reinforce public familiarity with Mars as a living world. (See Winter, "Early Spaceflight Simulation Shows," parts 1 and 2.) Image © Whipple Museum of the History of Science: Wh.6604; Wh.6605.

an appropriate author. Such a person would need to be both eminent enough to withstand the inevitable scrutiny that the article would elicit and reliable enough to give an account that was, in Newcomb's eyes, factual and cutting edge yet unsensational. Lowell complicated matters by personally contacting Newcomb with further comments and criticisms of his article for the tenth-edition supplements, against which the Washington astronomer maintained a thoroughly diplomatic stance. "I hope you fully understand the spirit in which I deal with the subject," he wrote to Lowell in May. "I have still a lack of absolute confidence in the complete objective reality of some of your conclusions; but I have no theory that I wish to sustain on the subject."[62]

These early preparations coincided with a returning wave of Martian sensation. The summer of 1905 brought the best opposition in six years, and the public once again turned its attention to the red planet. "A Yankee Circus on Mars" was the inaugural show at the New York Hippodrome in April, and in late May the front page of the *New York Times* carried the news that the Lowell Observatory had successfully photographed the planet's canals.[63] The paper had secured this scoop in the most direct manner possible, paying Lowell a fee in return for exclusive news from Flagstaff.[64] As with the *Herald* and William Pickering, the *New York Times* soon established a symbiotic relationship with Lowell, becoming both his news distribution agent and one of his fiercest promoters and defenders. "While Prof. Percival Lowell concentrates all his fine intellect and splendid energy on his investigations of the planet Mars," the paper wrote in 1907, "a large number of other astronomers and scientific writers seem to be devoting too much of their time to disproving his theories. This is a great waste of both time and intellect."[65]

Newcomb therefore tackled the problem of the Mars article at the same time that Lowell was regaining the public's attention. The photographs of the canals, in particular, garnered considerable press coverage, though in actuality their impact on the astronomical debate was mixed. As Jennifer Tucker has shown, the considerable problems associated with interpreting, let alone reproducing, the tiny, not very well-defined negatives taken at Flagstaff in 1905 and in Chile in 1907 gave the photos an ambiguous status as evidence. Most astronomers who saw the originals or the prints circulated with the Lowell Observatory Bulletin No. 21 found the canals difficult to make out beyond one or two dark bands, and they ultimately proved not to be the "doubt killing bullets from the planet of war" that Lowell trumpeted them to be.[66] As the astrophotography expert W. H. Wesley noted, "doubtless a photograph has no imagination, but imperfect definition applies equally to photographic and visual observations." H. H. Turner summed up the mood within the discipline when he wrote, "I do not mean to deny the claim any more than I am prepared to admit it. Personally I find it extremely difficult to say exactly what is on those prints, and friends to whom I have shown them differ a good deal in their interpretations." Nonetheless, major newspapers including the *Washington Post* and the London *Times*, taking their interpretations directly from Flagstaff, began to press the line that Lowell's photographs discredited the argument that the canals were mere optical illusions. Publicity of this sort was enough to worry Hale, who wrote to Barnard in November 1905 that "the facts are not generally known, since everyone will now be convinced of the reality of the canals on the supposition that photo-graphs cannot lie."[67]

Under these difficult circumstances, Newcomb approached Agnes Mary Clerke for advice on the issue of Mars and the *Britannica*. The two were friends, and Newcomb had recommended Clerke to Chisholm as a serious writer "without any of the rhetorical flourishes which pop-ular writers are so prone to indulge." Clerke proposed Edmund Ledger of London's Gresham College for the "Mars" article, describing him as "a sane thinker" and an "impartial collector of facts," but Newcomb appears not to have followed up on the suggestion.[68] By the autumn of 1905 he was, in fact, beginning to seriously consider the only astrono-mer he entirely trusted, writing to Chisholm that "it is not impossible that I may be willing to tackle the delicate subject of Mars myself."[69] In doing so, the question of impartiality had to be weighed up against the pressing matter of expertise. Newcomb was still an admitted neo-phyte on the subject, and he maintained in this period a spirited cor-respondence with Lowell on the latter's evidence and theories. Finally, on October 20 he came clean with his reason for carrying on such a discussion:

> I may add, in a confidential way, one reason of my interest in the subject is that an article on Mars must be prepared for the *Encyclo-paedia Britannica* within the next ten or twelve months, and that it desires to have some perfectly judicial authority well acquainted with all the circumstances prepare it. *It is desirable that this authority shall be one not engaged on either side of the questions which have arisen.* I fear that such an authority will be very difficult to find. On many accounts Maunder would strike me as a good man; but I fear that his experiments with the school boys a couple of years ago may stand in his way.[70]

Lowell was understandably appalled by the suggestion that Maunder might be given the job, emphasizing in a hurried reply to Newcomb that the Greenwich astronomer had a poor reputation and lacked ex-perience observing Mars. It was the very priorities that Newcomb gave greatest stock to, expertise and hard facts, that Lowell himself stressed in his response: "One has only to put himself in the reader's position to realize that *expertism* is as essential as impartiality and that the lat-ter quality is more possible with full knowledge of the subject than without it." What "the reader demands are the facts, all the facts and nothing but the facts." Newcomb in reply pressed at the question of whether such facts had been established by consensus: "It has been adopted as a part of the policy of the Encyclopaedia to avoid, so far as possible, the presentation of mere opinions or experiments or speculation as to what may be."[71] Clearly this was a matter of interpretation, and it was just this judgment over which Lowell's "expertism" sought to secure authority. Remaining in a quandary himself about the most suitable person to undertake the challenge of making these judgments, Newcomb invited Lowell to make his own suggestions as to the author. Bullishly, Lowell retorted that

> As to whom, in answer to your request, to recommend for its writer
> I find myself perplexed. On the one hand you wish the article well-
> written and on the other you wish no one who knows the subject
> well to write it; which is like asking for the best tables of the planets
> and expressly excluding those of any weight. You would be the first,
> and rightly, to deny validity to results based on such second-rate data;
> and precisely this will be the result of the article on Mars. I cannot
> think that any editor with his eyes open would want such when he
> is getting out a new edition on purpose to bring it up to date—
> especially when the sceptics have just been so signally defeated by
> photography.[72]

Lowell had a point, and being no closer to resolving this conundrum of expertise versus impartiality himself, Newcomb let the matter lie. The *Britannica* article would not go away, however, and in January 1906 the Washington astronomer confessed to his friend David Gill that "it is with many qualms of conscience that I have allowed myself to get as much interested in the [*Britannica*] work as I am—to the detriment of the completion of my astronomical work proper."[73] Three weeks later, at the point that Chisholm began to discuss the possibility of the canals' discoverer, Giovanni Schiaparelli, being given the Mars article, Newcomb took a decisive step toward a final decision. "I may have to undertake Mars myself," he wrote to Chisholm, "on the general principle of always being ready to take hold of what others might shrink from. I take it whoever prepares this article will have a squabble with Percival Lowell." A month later, relieved that Schiaparelli had not been approached, Newcomb made the decision official. He would write "Mars" himself, while attempting to "mitigate the inevitable fight with Lowell" by steering clear of "speculative ideas and fine drawn conclusions."[74] This would prove to be wishful thinking.

MARTIAN CANALS AND THE PSYCHOLOGY
OF PERCEPTION

Taking on the article meant that Newcomb would have to turn himself into an expert on Mars. Above all, this required him to get to grips with the work of the Lowell Observatory, the only major astronomical site dedicated to observing the planet continually. Extended correspondence with Lowell only took Newcomb so far in this quest, however, so from late 1905 he also began corresponding with a range of other experts who might aid his deepening research. Most significantly, these included a number of experimental psychologists, including Francis Galton, John W. Carr, and James McKeen Cattell, as well as one of Lowell's closest allies, the eminent New England zoologist Edward Morse.[75] Initially at least, it appears to have been the latter who had the most impact on the Washington astronomer's developing views. Director of the Peabody Museum in Salem, Massachusetts, Morse was an old friend of Lowell's and a well-respected member of the New England scientific elite. In May 1905 he had traveled to Flagstaff to spend five weeks observing Mars, an experience that convinced him of the veracity of Lowell's artificial canal hypothesis. This research and Lowell's theories were quickly assimilated into a book, *Mars and Its Mystery* (1906), a treatise on Martian planetology that capitalized on the author's scientific reputation to flesh out the planet's geological and biological life history. Morse was fiercely critical in the book of astronomers' "criticism or neglect" of Lowell's pioneering work, and he worked hard to promote his and Lowell's unique expertise as interdisciplinary interpreters of planetary markings.[76]

In November 1905, after likely being put up to the task by an increasingly worried Lowell, Morse wrote to Newcomb out of the blue. "I want to see you about Mars," he declared, before regaling his colleague with his experiences at Flagstaff. Newcomb was intrigued, but his response was mixed. Though he conceded to Morse the "enormous credit" due Lowell for the "energy and scientific spirit" of his work, he dwelt longest on the problem of Lowell's "opinionated" approach. Problems of speculation were intimately connected, in Newcomb's account, with matters of disciplinary hierarchy. Planetology certainly could not match up to his own positive astronomy. Lowell, he wrote, "talks as if the writing a description of Mars was as difficult and complex and required as rare a faculty as writing a book on celestial mechanics. What makes the matter worse is that . . . he combines with his observations proper so much doubtful speculation as to what is going on in Mars that it detracts from the high estimate which his observations justly command." But as Morse retorted, bold claims about the canal network were warranted precisely by that expertise: "If you realized the enormous amount of evidence [Lowell] has accumulated in the last few years with his great refractor . . . you would understand better the positive way in which he insists upon the presence of these markings." Pointedly, Morse suggested that it was only "those who are unfamiliar with the accumulated evidence" who took Lowell's approach to be dogmatic.[77] Expertise thus greatly problematized Newcomb's attempted subjugation of planetology's claims. Pertinaciously, in early December during a trip to Washington, Morse spent an evening as Newcomb's guest, showing him his drawings and manuscripts from Flagstaff. Newcomb was "flatfooted" on the subject, Morse warned him, and, just as Charles Darwin had "lived to witness the discomfiture of every one of his opponents, so it is with this theory of Mars and it requires very little prescience to foresee the time when one will wonder how it could ever have been disputed."[78]

Morse's impact on Newcomb's attitude to Lowell and Mars was dramatic. Newcomb explicitly conceded to Morse that their conversations had "materially alter[ed] the aspect of the case as I understand it," though he reiterated his imperviousness to the claim that intelligent action could be inferred from the appearances presented by the canals.[79] Soon after, the Washington astronomer discussed the matter further with his friend and fellow skeptic Gill, who turned down the chance to write "Mars" on the grounds that he could not condemn the canal theory without first visiting "those observatories where they say they can see the things which they drew." Newcomb's response was revealing:

> I think your views about Mars are very just from your point of view. Up to a recent time I have been myself very incredulous as to the nature of the canals. But, on looking more closely into the subject, I am coming around to believe not only in their actual existence which I never doubted when under a strict interpretation of the word "existence" but also in the substantial accuracy of Lowell's delineations. Especially interesting was a visit last summer of one of our eminent naturalists to the Lowell Observatory. During the first few nights he could see nothing of the canals but, after repeated trials, he learned how to recognize them and making his own independent drawings he found them to coincide with those of the other observers. I have

always been alive to the fact that results of this sort might be brought about by imagination and suggestion; and also to the possibility that by practice one may improve in seeing what is not there as well as in seeing what is there. But, in view of the fact that the facilities for observation are better at Flagstaff than at any other observatory; in view also of the long practice of the observers and in the partial confirmation of the principal canals by photographs I have about come around to accept Lowell's whole system in at least its main features.[80]

Newcomb's acceptance of the *Britannica* article therefore occasioned a complex reflection by him on the subject of Mars, its canals, and their interpretation. That the "repeated trials" of Morse brought him into closer agreement with the Lowell Observatory staff raised challenging questions about the primacy of experience in interpreting Martian features. Yet the possibility remained that Morse, like Lowell and his employees, might actually improve in seeing "what is not there" rather than what is.

This was a problem that linked astronomical expertise to one of the era's hottest scientific topics: the psychology of perception. In the United States, in particular, questions of the phenomenology of sense perception had become an abiding concern for a small cohort of scientist-philosophers interested in human consciousness and its relation to lived experience. By the early 1900s, this milieu had cohered into a philosophical movement—pragmatism—led by two of Newcomb's colleagues and correspondents, Charles Peirce and William James. Newcomb's growing struggles over the technical and social management of astronomical perception—including, ultimately, his attempts to solve the matter experimentally—must be understood in relation to these men's work and ideas. Writing "Mars" for the *Britannica* forced Newcomb to engage with a subject, psychology, that the pragmatists had investigated in considerable depth, and he did so in a way that engendered considerable antagonism between the two sides.[81] At its most basic level, this was because pragmatism was born of changes in American science and culture that Newcomb himself abhorred. The rapid and unsettling transformations of the fin de siècle—its strange forces and mysterious rays—fascinated the pragmatists, and from their study and their relations to human perception emerged a radical new set of ideas.

At the heart of these ideas was an antiformalist synthesis of philosophy and science that swept away notions of objective truth. Evaluating ideas with respect to their actual consequences, Peirce and James argued, begat a philosophical doctrine in which truth was nothing more than the sum of its practical results. On this telling, ideas are simply tools, knowledge is just an instrument for getting on in the world, and both are irreducibly social, saturated by the personal and social situations in which they are found. This was, in effect, the replacement of a positivist metaphysics with something like its opposite: a philosophy of consciousness and experience. One of pragmatism's major practical consequences therefore was the forging of a new discipline, the "new psychology," that worked at the junction of science and philosophy. Whereas early experimental psychology, most notably Wilhelm Wundt's work on the "personal equation" in astronomy, had associated the study of mental states with the measurement of sense data, the new psychology rejected such functionalism, looking beyond what people did to consider the ways in which they thought. As pioneers of this new hybrid practice, Peirce and James used laboratory techniques to experimentally probe the role of experience, in particular, in shaping human perception. The results were as equivocal as Wundt's had seemed clear-cut, and from them came an entirely new, antipositivist, vitalist, and sometimes mystical conception of human consciousness.[82]

James developed pragmatism in its mature form from 1898, crediting Peirce as the originator of the idea. He did so in part as a response against what he regarded to be the excessive scientism and materialism then prevalent in American culture, making Newcomb very close to the embodiment of everything that James was working to overthrow. If Newcomb believed in right and wrong as absolute terms, in truth as a definable entity, and in the necessary and complete separation of scientific thought and religious belief, then the "radical empiricism" of James's new psychology was predicated on its potential to transcend just these absolutes. "I believe," James wrote to a colleague, that "there is no source of deception in the investigation of nature which can compare with a fixed belief that certain kinds of phenomenon are *impossible*."[83] Pragmatism therefore nourished just the sorts of risky, speculative, and progressive science that men like Lowell pursued and men like Newcomb reviled. Clashes between the two camps were inevitable, and spoke to the rapidly growing divide in fundamental scientific methodology engendered by this new philosophy's rise.

Newcomb first clashed with James over the American Society for Psychical Research, founded in 1884 on the model of its London namesake, with Newcomb installed as first president. James had advocated research into psychical phenomena—not least in relation to the puzzling forces and rays of the "new" physics—as an antidote to the "passive disciples" of "closed and completed system[s] of truth."[84] Newcomb's election as president therefore seems curious, but it was a purely tactical move, intended to exploit the astronomer's famous skepticism. The choice was "an uncommon hit," James wrote to a friend—"if he believes, he will probably carry others."[85] This was a pretty big "if," and Newcomb didn't wait long before using his platform to direct a series of attacks against what he deemed to be the methodological shabbiness prevalent in the field. James responded in public and in private, finding fault with Newcomb's "a priori arguments" and his rigid inductive methodology, which he judged a poor fit for phenomena that, by their very nature, could not be experimentally controlled and replicated in the manner that Newcomb demanded.[86]

The well-founded suspicion was that Newcomb's extremely limited outlook on scientific practice betrayed an even more limited understanding of philosophy and thus of intellectual modernity. When Newcomb and Peirce clashed in the early 1890s over Peirce's (very pragmatic) definitions of "limit," "infinitesimals," and "doctrine of limits" for the Century Dictionary and Cyclopaedia (1889–1891), Newcomb eventually cut off the protracted debate with the declaration that "I have often been inclined to maintain . . . that all philosophical and logical discussion is useless. . . . It is pure nonsense to talk about one infinity being greater or less than another. . . . What more can I say?"[87] Peirce got his revenge privately in 1900 when he reviewed Newcomb's definitions for the Dictionary of Philosophy and Psychology (1901–1902), telling the work's editor that "Newcomb's articles are excessively bad. His notions of philosophy are such as an intellectual person about Cambridge, not a special student of philosophy, would pick up about 1855."[88] Newcomb's revenge was harsher yet. Unbeknownst to Peirce, Newcomb was instrumental in destroying his career, intervening twice to remove Peirce from the only two stable jobs he ever held. According to Peirce's biographer, Newcomb was motivated "by nothing more baleful than a sincere disgust for Peirce's morals and a jealous and ignorant misunderstanding of his work." Peirce died in abject poverty, a bitter and isolated man.[89]

It is within this ideological split that Newcomb's practical response to Lowell's arguments over the canals of Mars must be situated. Lowell was a personal friend of William James, and his own approach to investigations at the limits of scientific orthodoxy mirrored those of the

Harvard psychologist.[90] This divide had a practical as well as a philosophical dimension, which would prove to be particularly significant for Newcomb's growing interest in an *experimental* solution to the canal problem. At the same time that the pragmatists were founding a new psychology, Newcomb himself was actively engaged with related broad questions over how laboratory solutions might be brought to bear within proper scientific praxis. But whereas the pragmatists had used experiment to probe the problems and attributes of individuality, Newcomb saw experimentation as a means to eradicate it.[91] In Newcomb's philosophy of scientific method, meaning was derived entirely from measurement, and it was therefore a comprehensively Wundtian program that Newcomb hoped to bring to bear on the problem of canal observation. Newcomb already knew, after all, that it had been just such a method that had proven decisive in ameliorating the vexing problem of individual variation in positional astronomical observation. Faced with the fact that observers did not measure transit times consistently when compared to one other, observatory managers had triumphantly deployed electrotechnical and statistical tools to standardize measurements through the determination of each observer's "personal equation." Likewise, when attempts to secure a stable value for the astronomical unit through observations of the transits of Venus failed, Newcomb was instrumental in pushing for an experimental solution to the problem via a determination of the speed of light.[92] From this perspective, Newcomb could see experimental objectivity as a promising solution to a range of disciplinary maladies.

The challenge of drawing planetary detail was not directly comparable to the task of precisely timing the transit of a celestial object, however. Nevertheless, Newcomb clearly saw the problems as generally analogous, writing to Lowell that "I certainly would like very much to have a sort of competitive test and settle the matter. . . . I would bring you, Barnard and some of your observers together to examine various artificial objects more or less similar to the planets, without knowing what was marked on them to test the different qualities of vision."[93] There might be a "personal equation" for the perception of planetary detail, and from Newcomb's perspective the aim of experimental science was to obviate this problem rather than to explore or explain it. Unlike the pragmatists' more nuanced investigations, Newcomb therefore focused on the issue of fallibility. "The use and formation of mental images of cities, historical events, machines, mechanical forces, geometric constructions, arrangements of colors etc.," he wrote to Galton, "is really characteristic of the best trained and best educated men and comes largely from training and from education."[94] He reasoned, therefore, that if, below a certain definable level, even *expert* observers could be shown to mistake amorphous detail for linear features, then all discussion of a coherent canal network would be moot.

The question of expertise in planetary perception therefore needed to be investigated experimentally, and it would be the *Britannica* "Mars" article that pushed Newcomb to devise and carry out just such a test. Newcomb began composing the article in the second half of 1906, in the shadow of Lowell's *Mars and Its Canals* (1906). On January 7, 1907, Newcomb renewed his dialogue with the Flagstaff astronomer, asking him a series of technical questions about his observatory's optical practices, including about their use of color screens to restrict the spectrum of light observed and whether they had tried using a reflecting telescope rather than a refractor to eliminate the problem of chromatic aberration.[95] Then, on January 19, Newcomb wrote two significant and related letters. The first, to Chisholm, fretted over the inclusion of an illustration of Mars in the encyclopedia. Should they ask Lowell for a map, "leaving details to his judgement," or should the *Britannica* office try and make one themselves? On the same day, Newcomb set in motion his attempt to clarify exactly whether Lowell's judgment could be trusted. Writing in strictest confidence to his friend Edward Pickering, Newcomb explaining that he hoped, when in Boston in March, to "experiment on E. S. Morse and if possible on Lowell himself," as well as any other "experienced observers," to test their "visual inference." These experiments, which he hoped could be conducted at Harvard Observatory, would complement a paper he had already begun writing on "the optical principles bearing on the subject," intended for the *Astrophysical Journal*.[96] Pickering was game, but, perhaps predictably, Newcomb's hopes of snaring Morse or even Lowell were dashed when both men made excuses not to participate. Nevertheless, Newcomb went ahead, testing Solon Bailey and both Edward and William Pickering at Harvard and then later, at the instigation of the *Astrophysical Journal*'s editor Edwin Frost, E. E. Barnard and Philip Fox at the Lick.[97]

The anti-Lowell establishment were pleased with the results of these experiments, which soundly corroborated Newcomb and his allies' suspicions that trained observers were susceptible to the problem of erroneous visual inference (see figure 4.4). More experienced Martian observers appeared, in fact, to be *more* prone to draw artifactual linear detail, a result that Barnard dubbed "a splendid lesson."[98] In

PLATE I

A

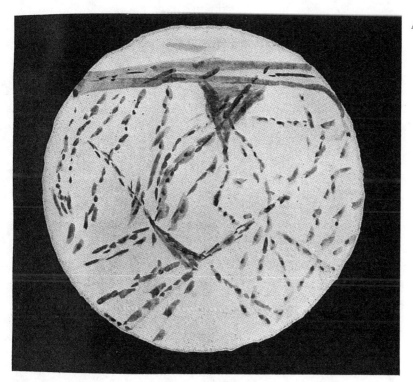

WASH DRAWING BY AUTHOR
Reduced from a diameter of 38 cm

Figure 4.4(A–B). Plates from Newcomb's "The Optical and Psychological Principles Involved in the Interpretation of the So-Called Canals of Mars," *Astrophysical Journal* 26 (July 1907): 1–17. A: the ink on paper drawing that Newcomb
used to test the astronomers. B: sketches by Solon Bailey and William Pickering
showing that, when observing at a distance giving the test image an equivalent
size to Mars seen through a powerful telescope at opposition, both men recorded
linear features that were not present on the test drawing.

early May, shortly before sending the finished article to press, Newcomb shared the gist of the results with Lowell, which were "very
striking in showing the extent to which the interpretation put upon
feint objects by practiced eyes was liable to deviate from reality." Tellingly, in the same letter Newcomb also asked Lowell to "look over the
draft of my proposed article on Mars for the *Britannica*." For both men,
these experiments were now intimately connected with Newcomb's
encyclopedia work, and as the Mars article neared completion they

PLATE II

SKETCHES

By S. I. Bailey   By W. H Pickering

B

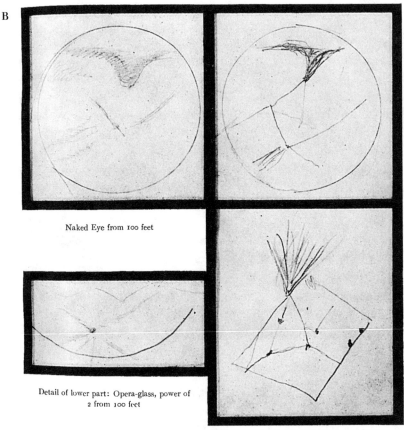

Naked Eye from 100 feet

Detail of lower part: Opera-glass, power of
2 from 100 feet

The upper Sketch with Naked Eye from 30 meters.
The lower, detail with opera-glass from 100 feet

began to argue vehemently over these intertwined projects. Newcomb now felt justified in restricting coverage of Lowell's conclusions only "so far as these conclusions relate to observed phenomena, and their explanation by known causes, a field to which the article is, by the rules of the Encyclopaedia, restricted." Lowell countered at length, breaking down the contrapositive conclusions of his own observatory's experiments upon the limits of perception before adding a long list of proposed corrections to Newcomb's draft article. For Lowell the stakes were clear: "As the *Encyclopaedia Britannica* is to last," his letter concluded, "do you not consider it but proper to quote at least so much?"[99]

AUTHORIAL POWER, EDITORIAL CONTROL, MEDIA AUTHORITY

In a complicated and at times turbulent debate over the validity of the Lowell Observatory's claims, the *Britannica* occupied a unique position. The "Mars" article, composed amid the height of another Martian sensation aroused by the oppositions of 1905 and 1907, became a central battleground for the establishment of disciplinary control over the narrative of an inhabited Mars. Yet because it was made "to last," its significance and role varied decidedly from that of the serial forms of publication analyzed in previous chapters of this book. For Newcomb, the opportunity to establish a rational response to Lowell was important enough that he took on the article himself, despite his own initial lack of expertise in the subject. With Lowell and his advocates so prominent in the ephemeral periodicals of the serial mass media, the *Britannica*'s reach and above all its permanence and authority offered an enticing opportunity for advancing counterarguments to a similarly massive audience. Newcomb's experiments into observational expertise and planetary perception are here shown to have arisen in direct response to this opportunity. Conversely, for Lowell and his allies, the importance of ensuring that their own expertise was fairly represented in the encyclopedia became crucial, provoking an extended campaign of pressure on Newcomb. Ultimately, for both sides the results were mixed.

This, in part, is because Newcomb's experimental campaign to discredit the existence of the canal network achieved only limited success. Lowell was provided space in the *Astrophysical Journal* to publish an extended response to Newcomb's paper, which he used to highlight both the Washington astronomer's lack of observational experience and his misunderstanding of astronomical optics. "If we put into our equations wrong fundamental facts," he wrote, "we shall only derive from it the wrong conclusions." Experience at the telescope, he countered, "and experiment there as well, are definite in their pronouncement against Professor Newcomb's supposed optical effects." Although Newcomb was given space for a short reply, Lowell in turn got a counter-reply, allowing him to get in the final word, which accused Newcomb—just as James had before him—of "a priori" arguments.[100] In the wider intellectual community, the impact of pragmatism certainly seems to have dampened enthusiasm for Newcomb's stridently positivist arguments. Barnard, one of Newcomb's key allies and test subjects, admitted in private to Lowell's colleague Vesto Slipher that "Newcomb's theoretical conclusions do not conform with practice." Peirce was even less kind, stating in a letter to Newcomb that "I do not think your investigation is up to your standard," as it "does not bring light upon the psychological laws involved."[101] Just as several facets of the canal debate—from photographic evidence to spectroscopic data to temperature calculations—appeared to hang in the balance, Lowell and Newcomb seemed to have fought to a competitive draw.[102]

An even worse fate was to befall Newcomb's *Britannica* article. A week after sending his visual inference paper to Frost, Newcomb sent "Mars" to Chisholm. Included in the piece was an account of his own experiments, as a counterpoint to a description of Lowell's canal network. Newcomb felt that he had been fair, advising Chisholm that "I found it unavoidable to relax a little our rule of omitting views and arguments and confining ourselves to well accepted facts. In this case it is impossible to draw a line between observed facts and inferences, and an article on Mars would be meagre indeed if it made no mention of mooted inferences." Chisholm, however, who had long seemed Newcomb's ally in the fight with Lowell, suddenly altered his stance, informing Newcomb that the article would be improved by "a short addition summarizing Lowell's own extended views and inferences." This addition was to be compiled by the *Britannica* staff and added at proof stage. Newcomb promptly protested, explaining that Lowell's principal inference, that the canals indicated the presence of intelligent action, was "a nearly impossible one, which a sane view can only regard as visionary."[103] But Chisholm stuck to his new plan, apparently swayed by the seeming weakness of Newcomb's own experiments and the extensive coverage then underway of the Lowell Observatory's successful expedition to photograph the canals from Chile.[104] Worse still for Newcomb, no opportunity to amend the altered proofs appears to have arisen. By early 1909 the venerable navy astronomer was suffering from the terminal stages of bowel cancer, and his archive contains no corrected proof sheets, suggesting that he died before he could comment on the final, much altered article. Had he seen the proofs, it would not have improved his health. Rather than assign to his staff the task of composing the Lowell material, Chisholm had actually approached Lowell himself, giving the Flagstaff astronomer space for two large planispheric maps of the canals and a 550-word footnote to Newcomb's article in which to outline his inhabited Mars theory at length.[105]

The resultant article was, consequently, something of a mess, apparently contradicting itself and providing no clear, "established" perspective on Mars.[106] In one respect, this can be attributed to the specific conditions under which it was produced. Developed on the lines of the new, cutting-edge, panoptic eleventh edition, and planned, written, reedited, and published during the height of the second great Mars boom of 1905–1911, the article clearly reflects the extraordinary uncertainty and conflict that characterized this period of Martian debate. But underneath this general current of uncertainty runs a much more specific explanation for the eventual shape of the article, which this chapter has traced in detail. We see in this account how, once again, the entwined practices of astronomers, authors, and editors combined with the resources and constraints provided by various genres of mass communication to shape the production of internationally distributed astronomical knowledge. The particulars of this story are distinct, however, from those that have preceded it in this book. With Proctor we traced a British contest over the moral status of astronomical speculation; with Pickering we followed a disciplinary fight over the control of transnational news distribution; with Newcomb we have seen how platforms of mass media outside the global news network were also integral to disputes over the use and status of experimental interventions in cutting-edge problems of planetary perception. This was an intra-disciplinary fight, with the *Britannica* not only facilitating but shaping Newcomb's intervention against the modish speculations of Lowell's neo-subdiscipline of planetology. Such an analysis therefore augments historical accounts of the place of new, purportedly objective observational and experimental practices in the physical sciences around the turn of the century, suggesting that the trials and tribulations of what might now seem self-evident practices can be explained by framing their emergence within a world shaped and bounded by its mass media.[107] The rise of disciplinary hierarchies controlled by professional associations and specialist journals runs parallel to, but also within, a much messier story of scientists who were also authors, lecturers, and editors, actively engaged with a world of public exhibition and mass-circulation print.

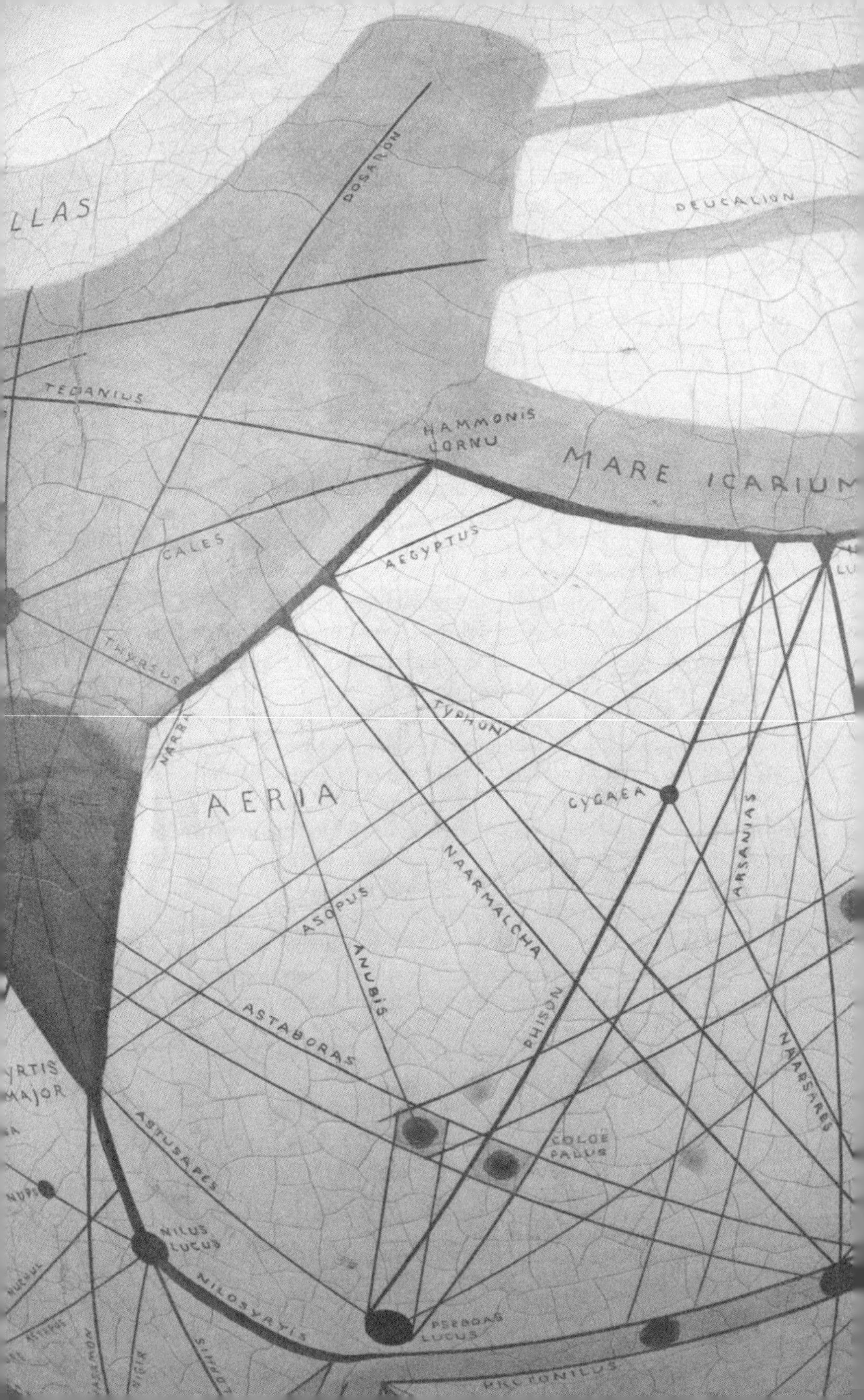

LLAS
DOSARON
DEUCALION
TEGANIUS
HAMMONIS CORNU
MARE ICARIUM
CALES
AEGYPTUS
THYRSUS
TYPHON
AERIA
CYGAEA
ARSANIAS
NAARMALCHA
ASOPUS
ANUBIS
PHISON
ASTABORAS
NAARSARES
YRTIS MAJOR
COLOE PALUS
ASTUSAPES
NILUS LUCUS
NILOSYRTIS
PSEBOAS LUCUS

CONCLUSION

In October 1916 Percival Lowell embarked on yet another of his famous lecture tours. Mars itself had not been in a good position to observe for nearly seven years, a fact that did not deter the Flagstaff astronomer from continuing to publicize the planet. Speaking before students on college and university campuses across the American West, Lowell remained bullish about his observatory's output, his complex theories of Martian life and society, and, crucially, the public's capacity to grasp and interpret these facts as educated consumers of Martian media:

> Twenty-two years ago we had proof sufficient to warrant the publication of the fact [of life on Mars]. . . . In the time that has elapsed further and further proof has been accumulated from long continued systematic observation of the planet. . . . Visual observation has stated the facts; photography has given its imprimatur to what observation disclosed; spectroscopy has yielded confirmation; and mathematical physics from testimony of its own has shown that we ought to see just what we perceive. . . . To fully appreciate the combined weight of this evidence it is necessary to consider it all, singly and collectively. . . . It must be read in its entirety and then anyone able to judge of evidence will see that its testimony is overwhelming.[1]

Lowell had good reason to be taking this argument directly to his most appreciative audience. As historians have chronicled in great

detail, the period immediately following his fight with Simon Newcomb over the *Britannica* was a difficult one for Lowell and his theory,
especially among his professional peers. The 1909 perihelic opposition
in particular had generated extremely skeptical reports from the select few observers in charge of the discipline's largest telescopes.[2] Although an aged Giovanni Schiaparelli had endorsed Lowell's ongoing
work shortly before his death, other allies were thin on the ground. W.
W. Campbell wrote to George Ellery Hale at Mount Wilson shortly
before the opposition to excoriate Lowell and his observing partner
David Todd: "You have of course noticed that Lowell, the past year or
two, has been making much ado in public, and in many matters quite
unprofessionally. . . . I think Lowell and Todd are going to be a trial
to sane astronomers. . . . My question is just how far they should be
allowed to go before somebody steps on their rope." Hale could only
agree, foregrounding in his reply the two camps' growing discord over
appropriate norms of discourse: "What I particularly dislike is [Low-
ell's] absolutely unscientific method of dealing with his material and
of stating his case."[3] Disagreement over evidence was inseparable from
disagreement over its mediation.

At the opposition itself, Hale made observations and photographs
with Mount Wilson's monster sixty-inch reflector that directly contradicted Lowell's. As a special cable from London to the *New York
Times* reported, these results fed into a new transatlantic alliance of
astronomers against the Flagstaff director. At a meeting of the British
Astronomical Association, Gresham Professor of Astronomy Samuel
Arthur Saunder had shown lantern slides of Hale's photos that, according to the telegram, the audience had considered an "enormous
improvement on any" yet seen. "The canals were not shown," Saunder
noted, perhaps because "the telescope was too strong to indicate them,"
a snide remark at Lowell's expense that had "evoked laughter" from
the audience.[4] At the same meeting, E. Walter Maunder read a report
by Eugène Michel Antoniadi that described Mars as "covered with a
vast and incredible amount of detail held steadily, all natural and logical, irregular and chequered, from which geometry was conspicuous by
its complete absence." As Maunder was pleased to report, these results
appeared to bring the idea of a "canal system" on Mars to "a most sat-
isfactory" conclusion. As the meeting's official write-up reported, both
the cause and effect of Martian sensation were now well understood by
the Greenwich astronomer:

There never was any real ground for supposing that in the markings observed upon Mars they had any evidence of artificial action. *Had it not been a sensational idea which lent itself to sensational writing in the daily press [Maunder] did not believe they would ever have heard of it.* He considered it was all the better for science that the idea was now completely disposed of. They need not occupy their minds with the idea that there were miraculous engineers at work on Mars, and they might sleep quietly in their beds without fear of invasion by the Martians after the fashion that Mr. H. G. Wells had so vividly described.[5]

Meetings like this and the widely circulated reports that followed constituted the first victory lap for a body of astronomers united by a fresh sense of purpose and shared disciplinary norms. This unity was forged in the fires of Martian scrutiny, and its result was a cohesive body of young, salaried observers in charge of giant telescopes and even larger budgets, who together abhorred everything that Lowell represented.[6] The vicious fight over control of the telegraphic distribution of planetary news that opened this book exemplifies this divide. But we need now to place it where it belongs, near the end of a story that is long and complex and that follows a path all too easily obscured by the crowing of what we now take to be its victors. Nor can this "end" be taken as sudden or definitive; there is strong evidence to suggest that from the perspective of the common reader or the general journal editor, Lowell's methods and arguments in no way appeared beyond the pale of mainstream science, nor did they suddenly cease to have valency in planetary science and its marketplaces (see figure C.1). Rather, the canals and ideas of intelligent life on Mars declined gradually, through what Robert Markley has described as "a complex and ambiguous process by which enthusiasm slid into agnosticism, and with interruptions, declined into skepticism."[7] The inconclusive and inconsistent nature of mass media like Newcomb's *Britannica* article provides one strand of evidence for such a nuanced process; the triumphant reports of Lowell's opponents, in contrast, certainly do not. When a reporter wired the Yerkes director Edwin Frost asking for "three hundred words expressing your ideas of the habitability of Mars," Frost jokes in his memoire that he sassily responded: "Three hundred words unnecessary—three enough—no one knows."[8] Lowell himself stayed at the eyepiece and in the public sphere for many years to fight against this rising tide of anti-canal sentiment from his colleagues

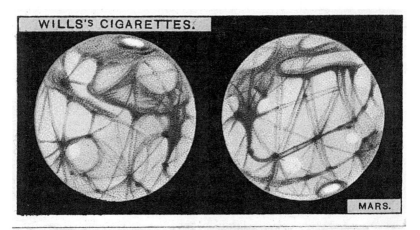

Figure C.1. Two cigarette cards from the 1923 Wills's Cigarettes series Romance of the Heavens. The bottom card shows an "Imaginary Landscape on Mars," illustrating "the theory of the late Prof. Lowell." The text on the card's reverse notes that "there has been great difference of opinion" over the validity of Lowell's work. Image © Whipple Museum of the History of Science, Cambridge: Wh.6067.

and peers, but the work took its toll. He died suddenly from a massive stroke in November 1916, only a month after his campus lecture tour had ended.

Forty-eight years later, he lay down on the couch for a session of psychoanalysis. His examiner, Charles K. Hofling, assistant professor of psychiatry at the University of Cincinnati, made a troubling diagnosis. Writing in the *British Journal of Medical Psychology*, Hofling argued that most of Lowell's major scientific claims were "heavily influenced by unconscious forces" that included "continuing anxiety toward [his] father" and "castration fear," which together were indicative of "unresolved oedipal conflicts." Such an analysis, Hofling stated, accounted for a range of Lowell's "sensory impressions and convictions" that "may, I think, be fairly categorized as illusory." Only through an examination of the long-dead astronomer's subconscious mind, he argued, was it possible to explain the "Martian fantasies" of Percival Lowell.[9] At the risk of venturing my own retrospective diagnosis, I would argue that Hofling's remarkable analysis actually reveals something quite different. With the benefit of another half century's distance, his paper now reads as a standout example of a more general malady, common among scientists and historians from the 1960s onwards and brought on by any talk of Martian canals. Its symptoms were characterized by a heady mix of incredulity, embarrassment, and forced explanation. Unable to come to terms with the extraordinary success, for at least four decades, of accounts of canals and theories of intelligent life on Mars, commentators from a range of disciplines began to look around for answers and excuses.

Astronomers were the hardest hit. "The canals of Mars," Carl Sagan and Paul Fox admitted, "have been something of an embarrassment to planetary astronomers." Unfortunately for them, cartographic science proved even less capable of providing concrete answers than psychoanalysis had. When the pair overlaid Mariner 9 photographs onto Lowell's maps, cross-comparison failed to demonstrate any correspondence between "the vast bulk of classical canals" and "the real Martian surface."[10] Little wonder, then, that others similarly troubled by the long shadow of the Mars controversy latched instead onto the triumphant reports of skeptics like Maunder, honing in on their characterization of the Martian features as nothing more than a translation error run amok in the popular press. "Had Lowell been less articulate, had he not directed his eloquence to the general public," Carl Sagan and Iosef Shklovskii lamented in 1966, "the [canals] debate would probably have terminated much earlier."[11] Unable to blame a simple trick of the eye, "popularization" became the favored stick to beat the canals with. Before long, a more or less fixed narrative had emerged and stabilized, recasting early planetary astronomers such as Richard Proctor, Camille Flammarion, William Pickering, and Lowell in the role of mere popularizers, capably exploiting a sensation-seeking press and an impressionable public in the face of professional opposition.[12]

Such accounts place media into the story of Mars in precisely the wrong way. At their most extreme, they posit popular media as a second-order phenomena, subsequent to and parasitic on scientific knowledge-making enterprises themselves. And even at their less severe, all Martian "popularization" narratives rely upon an asymmetrical and whiggish account that recycles the arguments they purport to explain. It is true that the contention of this book is that media is fundamentally implicated in the emergence and persistence of the claim that Mars was a living planet. But this is not a claim about mass media and Mars; it is a claim about mass media and astronomy. What I hope to have shown through this particular and limited set of case studies is that the emergent fields of astrophysics and (what we much later came to call) planetary astronomy were soaked through with the popular culture and social relations of Victorian and early twentieth-century mass media from their outset. It is, in fact, my contention that no aspect or output of these emerging fields can be understood accurately without incorporating such a frame of analysis, even though such a frame does not and cannot stand alone or outside many other strands crucial for a history of modern astronomy. This book therefore adds to rather than rewrites the wealth of existing historiography that already attends to these many other strands—be they technical, social, or cultural.

It is worth dwelling briefly on the ways in which this sort of reframing is itself historically contingent. The conclusions and lessons it helps us draw are not necessarily stable across the four case studies in this book, so we must remain sensitive to both commonalities and differences. Distinct to Proctor's identity, for example, was the claim that popular and financially successful scientific authorship should represent the principal arbiter of a practitioner's competence. This was an argument energized by its cultural and scientific moment: on the one hand, by conflicts over disciplinary identity in the face of novel astrophysical techniques, and, on the other hand, by conflicts over elitism and the place of knowledge and money in public and professional life. Our conclusions, then, are particular to this moment: that new journalism formed a constitutive part of the birth of astrophysics and that powerful alternatives to cloistered, hierarchical models for this discipline were in play at this time.

With the rise of the technological West came a completely new type of relationship between observatories and mass media. Especially as astronomy moved to the mountains, new global telegraphic networks constituted one key facet of a technology-driven discrimination in favor of specific, remote, sublime sites of privileged astronomical seeing. Planetary astronomy changed from a potentially placeless moral enterprise into a location-specific, high-cost, technological one. This broad assessment of the field from the 1880s onward engenders a commensurately wide-ranging conclusion: that the rise of the technological West incorporated a particular type of mediation within what is often taken as a purely disciplinary and financial triumph, the ascendancy of American astrophysics to world-leading status. These changes to the relationship between observatories and mass media in turn influenced how a new global marketplace for news from Mars understood and thought about that news, most strikingly in terms of long-distance signals and instantaneous messages.

The great Mars boom of 1892, as one key episode within this new global marketplace, illustrates how specific media technologies helped shape the emergence of particular kinds of knowledge about Mars. With mountain astronomers suddenly dependent upon transnational media networks to project their claims rapidly and widely, newspapers became the principal gatekeepers of Martian astronomy's information exchange. Tracing the serial progress of this new form of event astronomy vividly dramatizes the way globally networked mass media was implicated within the working practices of planetary observers in this era. Both the speed and the linguistic limits imposed by submarine telegraphy shaped and bounded what could be said about, and therefore what could be known about, Mars. The meaning and saliency of the "canals" themselves, I argue, emerged out of the spatio-temporally specific practices of this astronomical journalism.

The fallout of this remarkable episode reveals a need to revise and augment the most heavily trafficked era in the history of Martian astronomy. Percival Lowell, I argue, was not the only player at the table. But my analysis of encyclopedia and exposition work is also an attempt to recast the table itself—to draw our attention away from the "popular" and the "professional" by thinking instead about genre. Historians have long recognized that media choices within scientific praxis are one key component of boundary work. But, as Simon Newcomb's encyclopedia articles demonstrate, this work does not necessarily follow the top-down hierarchical trajectory that has typically been assumed. Even in the modern, post-Victorian era, a work that is more general, more widely read, more synoptic, more—yes—popular, is not necessarily further from the pen of a discipline's elite or from the coal face of scientific labor. Encyclopedias, which have an almost invisible profile in histories of the sciences after 1850, proved to be a powerful resource for astronomers keen to counter the immediacy of periodical literature with the permanence of massively popular works of reference. Looking closely, therefore, at how a single article in this genre was planned, commissioned, executed, and edited reveals a tellingly different story of what it meant to write public astronomy in the age of Lowell.

As distinct as these findings and conclusions are, they share a common orientation. They are all stories of media in science rather than science in media. That is, all of the main actors in this book, regardless of their roles and regardless of their views on Martian life, elide any formal distinction between what we might call "knowledge producers" and what we might call "media agents." As writers *and* astronomers, the relationships they had with mass media formed for all of them a constitutive part of their identities as practitioners. At least between the 1860s and the 1910s and across the astrophysical and planetary sciences, media relations were a constitutive part of astronomical practices, rather than serving as an adjunct to them. The problems of genre, speculation, and rhetorical form and the problems of disciplinary identity, institutional media relations, and astronomical technique were all inseparable. This is a claim that, of course, both draws upon and augments a broader argument now gathering momentum within the history of science, which considers science as a form of communication. The great majority of works that take this approach, however, focus their attention squarely on the marketplaces for scientific knowledge and on those markets' consumers.

My approach has differed insomuch as my abiding concern throughout this book has been on the work done to get knowledge to market. In this sense, I have written against the traditional frame of popularization studies, in which science is analyzed as a part of the public sphere. Here, I have argued for the reverse; by tracing the working practices of key practitioners across astrophysics and planetary science, with a focus on the media life-worlds they inhabited, I present the public sphere as an essential component of scientific practice itself. Of the many lessons that can be drawn from such a methodology, I especially want to stress one that has particular saliency for the story of news from Mars. It is a form of productive skepticism, not only of notions of diffusion *from* science *to* media but also of notions of negotiation *between* science and media. Such ideas of conference and exchange risk presupposing separations and hierarchies among actors and roles, and they carry with them an assumption that this book works to oppose:

that media expertise and scientific expertise are inherently distinct skillsets, embodied by distinct practitioners.

An extension of this form of skepticism might be profitable in a wide range of subject areas and time periods. Much recent scholarship has already reinforced the notion that careful attention to genre and rhetorical form can reveal crucial features of the sciences' past missed by disciplinary histories. Recovering media in scientific praxis promises to fruitfully extend these antidisciplinary narratives while enabling us to probe the changing place of media qua praxis. The era immediately following Lowell appears particularly ripe for such scrutiny. After all, it seems clear, at first blush, that Lowell's rhetoric of public knowledge-making—of evidential judgment through and by mass media—became antithetical to the practices of a freshly minted "planetary science." This is a change typically glossed by the catchall concept of "professionalization," as embodied by Lowell's high-status critics. Yet the findings of this book suggest that we cannot at all assume so simple, sudden, or explicable a transition away from a powerful and entrenched mass-media-centered system of information exchange and knowledge-making. Indeed, what scant historical analysis we have of this period would suggest just the opposite: Lowell's canals remained much discussed, the idea of a living but dying planet remained prominent in astronomical discourse, and the use and abuse of contested media genres and formats continued to be a battleground for the discipline.[13] Rather than assuming the canals' demise at the hands of professionals, then, the particular disciplinary transformations that eventually made their demise seem so self-evident need to be carefully explained. Such an approach promises, in turn, to give us a much richer understanding of the wider dynamics of disciplinarity and public science in the interwar era—of how stable roles for academic scientists and science correspondents came into being, or didn't—and of how astronomy was, or wasn't, implicated in the emergence of a thoroughly modern assumption of a "gap" between science proper and its passively consuming publics.[14]

Among the many potential topics ripe for such an analysis, I highlight two as illustrative. The first is the "new physics" of Einsteinian relativity and quantum theory around and after 1900. Superficially, at least, the radical changes to the scope and practices of physics in this era appear to echo quite closely the disciplinary transformations of the new astronomy that began a generation earlier. Recent accounts of the "race" to test relativity indicate clearly how international networks of telegraphic exchange played a central role in these trials.[15] It remains the case, however, that such accounts risk treating these seemingly straightforward transmission systems as little more than frictionless aids to rapid information circulation. As such, there appears to be further scope for considering the means by which mass media technologies were implicated within the working worlds of early twentieth-century physics and the consequences of this entanglement. In particular, a range of recent studies that have focused on the reception of general relativity have already begun to bring into focus the significant role played by newspapers and popular authorship in this process. Accounts of the public's understanding of Einstein's work, of the co-construction of "classical" and "modern" physics, and of popular reports substantiating evidence for the relativity "revolution" all point to the ways in which mass media formed a constituent part of physics' disciplinary landscape.[16] Such accounts are neither exceptional nor aberrant. Rather, they might represent the first steps in a comprehensive reassessment of new physics as an emergent discipline co-constructed with and through the cultural marketplaces of fin-de-siècle mass media.

The second obvious place to deploy and extend this book's analytical model would be to return to Mars, to the perihelic opposition of 1924, the first since 1909. The events of that year have received strikingly little attention in disciplinary accounts, presumably because the implication of mass media in much of this work seems so alien. On the terms of planetary science, it can be hard to know what to make of the U.S. Army's and Navy's radio stations being ordered silent for coordinated periods across three days during the opposition to facilitate detection of broadcasts from Mars. The results, even more so: only days into the enterprise, mysterious dots and dashes were "reported heard at the same time by widely separated operators of powerful stations," with David Todd, professor emeritus at Amherst College, quickly deploying a "radio photo message continuous transmission machine" to record the signals onto film. The chiefs of the code section of the army's signal office and the radio division of the Bureau of Standards were soon drafted in to help decipher these recordings, sparking a very public debate over whether the photographic record of jumbled dots showed "a crudely drawn face," or merely terrestrial radio interference. "Possibly," wrote one correspondent to a Boston paper, "there is nothing left of the inhabitants of Mars except heads." Across the Atlantic, astronomers in the Swiss Alps and France attempted to find out, signaling back with light messages and radio waves.[17] This book has argued that we cannot make sense of episodes like these through accounts that consider them as merely media events; nor can we understand them by attempting a Procrustean framing within disciplinary histories of modern planetary science. Rather, we will need instead to assess them as episodes of media in astronomical praxis. Only on these terms can we explore the complex, contingent ways in which astronomy's diverse cultural marketplaces continued—and continue—to influence the production of news from Mars.

ACKNOWLEDGMENTS

As befits my day job, I will start by thanking an object. I first landed on Mars as a research topic thanks to the remarkable globe that adorns this book's cover. Its arrival at the Whipple Museum of the History of Science a decade ago introduced me to the idea of an inhabited Mars, and I am indebted to the museum's director and curator, Liba Taub, for encouraging me to research this extraordinary object. In developing those initial tentative inquiries into this book, I owe a debt above all to three brilliant and patient guides: Jim Secord, Simon Schaffer, and Liba. Some years later, I'm privileged to still draw on the wisdom and support of these three mentors, who are now also friends and colleagues.

In the long journey across time and space between my first encounter with life on Mars and the completion of this book, I've received invaluable feedback from a number of other scholars kind enough to read and critique my work. I thank especially Helen Curry, David DeVorkin, Alexandra Ion, David Singerman, two anonymous referees for the journal *Isis*, and two anonymous referees for the University of Pittsburgh Press. Special thanks go to Mary Brazelton, whose careful and astute reading of my manuscript got me through the most difficult phase of its completion and made this a much better book. Thanks too to the many other scholars who have conversed with me and shared ideas along the way, especially Jenny Bangham, Robert Bud, Jenny

Bulstrode, Steven Dick, David Hecht, Ruth Horry, Jeff Hughes, James Hyslop, Boris Jardine, John Jones, Nicky Reeves, and Richard Staley.

Bernie Lightman, as well as being a brilliant and generous reader, has been an immensely patient editor. I thank him for the invitation to contribute to the Science and Culture in the Nineteenth Century series and for his unceasing encouragement and support. Thanks also to the University of Pittsburgh Press for producing this book, and especially to Abby Collier for making the review and revision process so painless and for fielding my steady stream of wide-eyed questions and queries.

The Cambridge Department of History and Philosophy of Science has been a wonderfully supportive and stimulating environment in which to work. For making it such an enjoyable and accommodating place, special thanks go to Toby Bryant, Lukasz Hernik, Tamara Hug, Agnieszka Lanucha, James Livesey, Mark Rogers, Louisa Russell, and David Thompson. In the Whipple Library, Jack Dixon, Tim Eggington, Anna Jones, and Dawn Kingham have all been unceasingly helpful. In the Whipple Museum, I've had the pleasure of researching and working alongside a fantastic team who together have put up with years of Martian invasions and interruptions. My heartfelt thanks to Lorena Blythe, Rosanna Evans, Steve Kruse, Alison Smith, and Claire Wallace.

It would not be possible to list all the librarians, curators, and archivists who have assisted me on Mars, but special thanks must be extended to the following: Mark Hurn, librarian at the Institute of Astronomy, Cambridge; the staff in the manuscripts reading room of Cambridge University Library, in particular Zoe Rees; Kathryn McKee of Saint John's College Library, Cambridge; Dick Chambers at the British Astronomical Association; Angela Mandrioli at Exeter University Library Special Collections; and Dan Mitchell at University College London Special Collections. In the United States, I was fortunate enough to spend a deeply rewarding research stay at the John W. Kluge Center in the Library of Congress. I thank David DeVorkin and Deborah Warner for their generous support and hospitality and Mary Lou Reker, Carolyn Brown, Alisha Robertson, and all of the staff and scholars at the center who contributed to its rich intellectual environment. I thank also Tom Mann, everyone at the Manuscript Reading Room of the Library of Congress, everyone at Harvard University Archives, Kit Messick of the Manuscripts and Archives Division of the New York Public Library, and Tom Tryniski, whose Old Fulton NY online newspaper collection is an extraordinary and invaluable resource for the historian of American mass media.

I am extremely grateful to the Arts and Humanities Research Council for funding much of my research, including my Fellowship at the John W. Kluge Center. I thank also Trinity Hall, the Department of the History of Philosophy of Science, and the British Society for the History of Science for providing grants to facilitate research and conference participation.

Finally, a huge thanks to all my friends and family. Special thanks to friends, near and far, who've supported me on my Martian travels: Mary, John, Sinead, Rob, Toby, Dan, Luke, Neil, David, Mary, Boris, Jenny, James, Ruth, Steve, Helen, Andrew, Julia, Kat, Chris, Elis, and John. And, above all, thanks to Dennis and Ingeborg, Colin, Isabelle, Pedro and Miguel, Kellogg and Diana, Sam and Sarah, Mum and Dad. This book was born at Mount Pleasant Barn in a home packed with books, questions, love, and a deep commitment to learning. Here's my small thanks for the education.

NOTES

ABBREVIATIONS

HUA-ECP   Harvard College Observatory, Records of Director Edward C. Pickering, 1864–1926. HOLLIS No. 001966647. Harvard University Archives, Cambridge, MA.

HUA-WHP   Papers of William H. Pickering, 1868–1916. HOLLIS No. 000604354. Harvard University Archives, Cambridge, MA.

LOA-ECLO   Early Correspondence of the Lowell Observatory, 1894–1916. Microfilm Edition. Lowell Observatory Archive, Flagstaff, AZ.

LOC-SN   Simon Newcomb Papers, 1854–1936. MSS34629. Library of Congress, Washington, DC.

NYPL-HMAPD   Henry and Mary Anna Palmer Draper Papers, 1859–1914. MssCol 838. New York Public Library, New York.

SJCL-JCA   Papers of John Couch Adams. GBR/0275/Adams. Saint John's College Library, Cambridge, UK.

SJCL-RAP   Letters of Richard Anthony Proctor. PR2—Miscellaneous Papers. Saint John's College Library, Cambridge, UK.

UCL-FG   Papers and Correspondence of Sir Francis Galton. GB 0103 GALTON. University College Library, Special Collections, London, UK.

UEL-NL   Sir Norman Lockyer Research Papers. EUL MS 110. Special Collections, University of Exeter Library, Exeter, UK.

INTRODUCTION

1. Jones and Boyd, *Harvard College Observatory*, 194–98.

2. "Lowell to Give Out Planetary News," *Kansas City Times*, Nov. 3, 1909, 9; Strauss, *Percival Lowell*, 234–37.

3. "Harvard Observatory Still News Center Says Professor Pickering," *Boston Journal*, Nov. 6, 1909, 7.

4. Clerke, *Popular History of Astronomy*, v. Owen Gingerich, in his *General History of Astronomy*, vol. 4, 3, gives a helpful definition of astrophysics as "physical studies of celestial objects." However, modern understanding of this field tends to limit its scope somewhat, placing the study of the topography and life history of a planet into a separate domain—"planetary science." Yet the hiving off of "planetary science" postdates the timeframe of this book. As such, I follow the habits of my actors in thinking about astrophysics as more or less interchangeable with the concept of "new astronomy," that is, a field that incorporated *all* aspects of the physical study of planets and other objects, including visual work and mapping.

5. The first quote is from Robert Bunsen and Gustav Kirchhoff, "Chemical Analysis by Spectrum-observations," *Philosophical Magazine*, 4th ser., 20 (Aug. 1860): 89–109, 89. The London astronomer was Warren De la Rue, speaking before the Chemical Society in 1861, as quoted in Becker, *Unravelling Starlight*, 22.

6. Clerke, *Popular History of Astronomy*, 2. On the early history of astrophysics, see also Becker, *Unravelling Starlight*; Gingerich, *General History of Astronomy*; Hufbauer, *Exploring the Sun*, 42–70; Bigg, "Staging the Heavens"; Brück, *Agnes Mary Clerke*; Schaffer, "Where Experiments End"; Hearnshaw, *Analysis of Starlight*, 21–62; and the two-part special issue on "Spectroscope Histories" in *Nuncius* 17, no. 2 (2002): 583–689 and 18, no. 2 (2003): 737–852.

7. Clerke, *Popular History of Astronomy*, v, 6.

8. Secord, *Victorian Sensation*; Secord, *Visions of Science*; Fyfe and Lightman, *Science in the Marketplace*; Morus, "Manufacturing Nature"; Topham, "Scientific Publishing."

9. Wiener, *Americanization*, 1–27. The attack from Arnot Reid is quoted on 18–19; the endorsement by W. T. Stead is quoted on 20.

10. I draw on a variety of works that have assessed the shifting centers of orthodoxy and heterodoxy in Victorian science. Useful overviews are given in Winter, "Construction of Orthodoxies and Heterodoxies"; and Clifford et al., *Repositioning Victorian Sciences*. Specific studies of phrenology, evolutionary theories, mesmerism, and spiritualism are particularly relevant. See Cooter, *Cultural Meaning of Popular Science*; Shapin, "Politics of Observation"; Secord, *Victorian Sensation*; Winter, *Mesmerized*; Noakes, "Spiritualism, Science and the Supernatural"; Oppenheim, *Other World*. Work in the sociology of scientific knowledge demonstrates the need to assess controversies symmetrically, since the closure of debates tends

to efface the work through which the norms and boundaries of the victors were established. See Collins, *Changing Order*, 5–28; Latour, *Science in Action*, 96–100, 141–44; Shapin and Schaffer, *Leviathan and the Air-Pump*, 3–7.

11. Bensaude-Vincent, "Science and the Public."

12. Much of this overview is drawn from the best general history of Martian studies, William Sheehan's *The Planet Mars*. A more detailed history of Martian mapping is given in Lane, *Geographies of Mars*, 23–63.

13. Crowe, *Extraterrestrial Life Debate*, 167–355. Crowe's quantitative assessment (351–52) of the notorious 1850s fight between David Brewster and William Whewell finds about two-thirds of published works supported the pluralist Brewster against the antipluralist Whewell. See also Brooke, "Natural Theology."

14. Huggins's results greatly overestimated the presence of water in the Martian atmosphere, a fact that would not become clear for many decades. See DeVorkin, "W. W. Campbell's Spectroscopic Study."

15. Proctor, *Other Worlds Than Ours*, xiii, 57. Proctor's work and methodologies are analyzed extensively in ch. 1 herein. For an overview, see also Crowe, *Extraterrestrial Life Debate*, 367–77.

16. The periodic cycle of close approaches of Mars to Earth is discussed further in the next section of this introduction and in figure I.2.

17. It is at this point that most accounts of the "Martian canal controversy" attribute its root cause to an unwitting mistranslation from Schiaparelli's "canali" (channel *or* canal in Italian) to "canal" in English. This explanation is, I believe, entirely unsatisfactory. Lane, *Geographies of Mars*, 24, provides a persuasive image-based explanation for why we should not take the mistranslation claim seriously. In ch. 3 I offer a further mass-media-based explanation.

18. Lane, *Geographies of Mars*, 23–63, 44.

19. For examples, see n9 and nn11–12 in the conclusion herein.

20. This literature is reviewed in both Strauss, "Reflections"; and Smith, "Martians and Other Aliens." The major works are Hoyt, *Lowell and Mars*; DeVorkin, "W. W. Campbell's Spectroscopic Study"; Burnett, "British Studies of Mars"; Webb, "Planet Mars and Science"; Webb, *Tree Rings and Telescopes*, 37–53; Crowe, *Extraterrestrial Life Debate*, 480–546; Sheehan, *Planets and Perception*; Sheehan, *Immortal Fire Within*, 237–63; Sheehan, *Planet Mars*, 58–145; Osterbrock, "To Climb the Highest Mountain"; Plotkin, "William H. Pickering in Jamaica"; Strauss, "'Fireflies Flashing in Unison'"; Strauss, "Founding of the Lowell Observatory"; Putnam, *Explorers of Mars Hill*, 3–42; Dick, *Biological Universe*, 62–105.

21. An important exception to this lack of analysis of Lowell's opponents is DeVorkin, "W. W. Campbell's Spectroscopic Study." Osterbrock, "To Climb the Highest Mountain," also addresses a similar story, but it does so while making the sweeping and questionable claim that "the professional astronomers of his time

never took Lowell seriously. To them he was a rank amateur who did not understand the methods of science" (80).

22. This literature is briefly surveyed in Nall, review of *Geographies of Mars*. The major works are Lightman, "Visual Theology of Victorian Popularizers," 661–71; Strauss, *Percival Lowell*; Markley, *Dying Planet*, 1–149; Tucker, *Nature Exposed*, 207–33; Lane, "Geographers of Mars"; Lane, "Mapping the Mars Canal Mania"; Lane, *Geographies of Mars*; Canadelli, "'Some Curious Drawings'"; Crossley, "Percival Lowell"; Crossley, *Imagining Mars*; Willis, *Vision, Science and Literature*, 57–113.

23. Works that assess pre–Space Age Mars in science fiction include Markley, *Dying Planet*, 115–49; Willis, *Vision, Science and Literature*, 89–113; Crossley, *Imagining Mars*; Crossley, "H. G. Wells, Visionary Telescopes"; Crossley, "Mars and the Paranormal"; Guthke, *Last Frontier*, 324–92; Tattersdill, *Science, Fiction*, 28–61; Fayter, "Strange New Worlds"; Mullen, "Undisciplined Imagination"; Schroeder, "Message from Mars"; Hendrix, Slusser, and Rabkin, *Visions of Mars*; McLean, *Early Fiction*, 89–113; Bould and Miéville, *Red Planets*.

24. Porter, "Gentleman and Geology"; Secord, *Victorian Sensation*, 524; Golinski, *Experimental Self*, 3–4; Barton, "'Men of Science'"; Desmond, "Redefining the X Axis"; Alberti, "Amateurs and Professionals."

25. The most forceful case for this bifurcation can be found in works on the history of American astronomy that treat changes within the discipline after 1850 as a "triumph of professionalism" over inferior, dilettante "amateurs." See, for example, Lankford, "Amateurs versus Professionals," quote on 11; Lankford, "Amateur versus Professional"; Lankford, "Amateurs and Astrophysics"; Lankford, "Astronomy's Enduring Resource"; Hetherington, "Amateurs versus Professionals"; Rothenberg, "Organization and Control." I am more persuaded by the nuanced and skeptical analyses of the progress and problems associated with professionalization in nineteenth-century American science offered by Reingold, "Definitions and Speculations"; and Lucier, "Professional and the Scientist." In the case of astronomy, the tag of "amateur" better fits the sort of wealthy "Grand Amateur" practitioners working earlier in the century, as analyzed in Chapman, *Victorian Amateur Astronomer*.

26. Useful overviews of this field and its methodologies include Cooter and Pumphrey, "Separate Spheres and Public Places"; Secord, "Knowledge in Transit"; Secord, "Electronic Harvest"; Topham, "Scientific Publishing"; Topham, "Rethinking the History"; Topham, introduction to *Isis* Focus section, "Historicizing 'Popular Science'"; Shapin, "Science and the Public"; Fyfe and Lightman, *Science in the Marketplace*; Knight, "Scientists and Their Publics"; Hilgartner, "Dominant View of Popularization." Some example works, from a now-extensive literature, include Lightman, *Victorian Popularizers of Science*; Secord, *Victorian Sensation*; Fyfe, *Science and Salvation*; Morus, "Manufacturing Nature"; Mussell, *Science,*

Time, and Space; O'Connor, *Earth on Show*; Keene, "Familiar Science"; and the products of the "Science in the Nineteenth-Century Periodical" project: Cantor et al., *Science in the Nineteenth-Century Periodical*; Cantor and Shuttleworth, *Science Serialized*; Henson et al., *Culture and Science*.

27. Bensaude-Vincent, "Science and the Public." Collini, *Public Moralists*, 199–250, offers a cogent critique of simple "professionalization" narratives in this era.

28. This need for specificity is noted, for example, by Bensaude-Vincent, "Science and Its 'Others,'" 360–61. On "spectacular science," see, for example, the special issue on "Spectacular Astronomy" in *Early Popular Visual Culture* 15, no. 2 (2017); Morus, "Worlds of Wonder."

29. Notions of "identity formation," "self-fashioning," and "boundary work" have all been influential on my approach to the relations of media and scientific personhood. See especially Golinski, *Experimental Self*; Barton, "'Men of Science'"; White, *Thomas Huxley*; Ellis, *Masculinity and Science in Britain*; Sponsel, *Darwin's Evolving Identity*; Morus, "Different Experimental Lives"; Aubin and Bigg, "Neither Genius nor Context Incarnate"; Jardine, "Between the Beagle and the Barnacle;" Shapin, "Image of the Man of Science"; Thurs, *Science Talk*; Gieryn, "Boundary Work."

30. Secord, "Knowledge in Transit"; Lewenstein, "From Fax to Facts"; Mussell, *Science, Time, and Space*; Secord, *Victorian Sensation*; O'Connor, *Earth on Show*; O'Connor, "Reflections on Popular Science"; Taub, *Science Writing in Greco-Roman Antiquity*, 1–21; Tattersdill, *Science, Fiction*, 1–27; Frasca-Spada and Jardine, *Books and the Sciences in History*; Hochadel, "Making of a Magic Mountain."

31. Useful audience-focused accounts of the rise of mass media in the second half of the nineteenth century and consequent defenses of the use of the word "mass" in cultural studies are given by Carey, *Intellectuals and the Masses*; Carey, *Communication as Culture*; Rydell and Kroes, *Buffalo Bill in Bologna*, 1–13; Wiener, *Americanization*; Broks, *Media Science*, 1–14.

32. Topham, "Rethinking the History."

33. The idea of a "mediatized" science is I suspect an entirely twentieth-century one. See, for example, Carson, *Heisenberg in the Atomic Age*, 114, for the assumption that scientific events are "mediatized" because "someone took care to contact the media, and someone sent journalists to record and report."

34. Hopwood, Schaffer, and Secord, "Seriality and Scientific Objects," 261. See also Mussell, *Science, Time, and Space*.

35. Carey, *Communication as Culture*, 155–77; Marvin, *When Old Technologies Were New*; Colligan and Linley, "Nineteenth-Century Invention of Media"; Kern, *Culture of Time and Space*; Darnton, "Early Information Society."

36. Wiener, *Americanization*, 5.

37. Gordin, *Scientific Babel*, 7, notes that scholarly publication at the turn of

the twentieth century was still at least trilingual, being dominated by a "triumvirate" of languages—English, French, and German—each with a roughly equal market share. On Flammarion's delayed impact on transatlantic news from Mars, see ch. 1, n135 herein. For examples of his work for the *New York Herald*, see chs. 2 and 3 herein.

ONE: Writing on Mars

1. W. Noble, "Richard Anthony Proctor," *Observatory* 11 (Oct. 1888): 366–68, 366.

2. Webb, letter to the editor, "Canals on the Planet Mars," *Times* (London), Apr. 10, 1882, 4. For an explanation of the concept "perihelic opposition," see figure I.2.

3. Proctor, letter to the editor, "Canals on the Planet Mars," *Times* (London), Apr. 13, 1882, 12. On the significance attributed to this letter in recent accounts of Mars and its canals, see, for example, its reproduction in Colin Pillinger's popular account of Martian exploration, *Beagle*, 44.

4. Dick, *Plurality of Worlds*; Dick, *Biological Universe*; Crowe, *Extraterrestrial Life Debate*.

5. Crowe, "Proctor, Richard Anthony," 417. On Proctor's huge appeal and popular impact, see also Lightman, "Astronomy for the People"; Lightman, "Visual Theology of Victorian Popularizers"; Lightman, *Victorian Popularizers of Science*, 295–351.

6. Schaffer, "Where Experiments End," 267–76. Huggins's reminiscences, given in an 1897 essay fittingly entitled "The New Astronomy: A Personal Retrospect," are quoted on 268.

7. Huggins and Miller, "On the Spectra of Some of the Fixed Stars," *Philosophical Transactions of the Royal Society of London* 154 (1864): 413–35, 434. On the contribution of spectroscopic evidence to pluralist debates, see pt. 3 of Crowe, *Extraterrestrial Life Debate*.

8. This disciplinary instability is discussed in greater detail later in this chapter.

9. Lightman, *Victorian Popularizers of Science*, 300–307.

10. Lightman, "Popularizers, Participation," 347.

11. The review appeared in the popular generalist science periodical *Scientific Opinion*, as quoted in Lightman, "Popularizers, Participation," 347. Proctor was far from the only science worker to fashion an identity that straddled esoteric and exoteric realms. See, for example, Rankin and Barton, "Tyndall, Lewes."

12. Hollis, "Decade 1870–1880," 211. See also Williams, "Astronomy in London"; Chapman, *Victorian Amateur Astronomer*, 113–44.

13. On the Greenwich-Cambridge axis and the dominance of Airy, see Dewhirst, "Greenwich-Cambridge Axis"; Chapman, "Private Research and Public Duty"; Smith, "Cambridge Network in Action"; Smith, "National Observatory

Transformed." On the RAS, positional astronomy, and "business astronomers," see Ashworth, "Calculating Eye"; Ashworth, "John Herschel, George Airy."

14. Smyth in his *Sidereal Chromatics*, as quoted in Schaffer, "Where Experiments End," 263.

15. Proctor, *Rough Ways Made Smooth*, 77.

16. On "gentlemanliness," "men of science" and changing patterns of Victorian scientific decorum, see White, *Thomas Huxley*; White, "Conduct of Belief"; Morrell and Thackray, *Gentlemen of Science*; Endersby, "Odd Man Out." On the X Club and professionalization, see Barton, "Influential Set of Chaps"; Barton, "Huxley, Lubbock"; Barton, "'Men of Science'"; Jensen, "X Club"; MacLeod, "X Club." On Victorian Scientific Naturalism, see Dawson and Lightman, *Victorian Scientific Naturalism*; Lightman and Reidy, *Age of Scientific Naturalism*; Lightman, "Science and the Public"; Lightman, "Creed of Science"; Desmond, *Thomas Huxley*.

17. Lightman, *Victorian Popularizers of Science*, 318–25.

18. Meadows, *Science and Controversy*, 96.

19. Ratcliff, *Transit of Venus Enterprise*, 46–56, provides an excellent account of this episode. The *Spectator* article is quoted on 49. See also Hollis, "Decade 1870–1880," 178–85.

20. Proctor, *Universe and the Coming Transits*; Proctor, *Transits of Venus*. Proctor also wrote various articles on the matter for popular periodicals, which were typically recycled in one of Proctor's many books of collected essays. See, for example, Proctor, *Essays on Astronomy*, 372–401.

21. Quoted in Ratcliff, *Transit of Venus Enterprise*, 51.

22. Proctor to Airy, Jun. 26, 1872, as quoted in Meadows, *Science and Controversy*, 98. Different aspects of this controversy are also covered in Hollis, "Decade 1870–1880"; Becker, *Unravelling Starlight*, 136–45; Macdonald, *Kew Observatory*, 143–45; Williams, "Astronomy in London," 20–22; Ratcliff, *Transit of Venus Enterprise*, 52–54. Proctor did not keep his views private. In his *Wages and Wants*, 6, for example, he wrote that "so rife has jobbery been of late in certain quarters, that some have not been ashamed to advocate a proposal for placing this person [Lockyer], ignorant of the very elements of astronomy, at the head of an important proposed astronomical observatory."

23. Lockyer to Airy, Nov. 7, 1872, as quoted in Meadows, *Science and Controversy*, 99.

24. Proctor's allies included John Browning, T. W. Burr, E. B. Dennison, and William Noble.

25. Airy to Adams, Jan. 6, 1873, Papers of John Couch Adams, GBR/0275/Adams, Saint John's College Library, Cambridge, UK, hereafter cited as SJCL-JCA (emphasis mine). Airy's response to Adams's enquiry about Proctor included a copy of a letter he had already sent to Arthur Cayley, president of the society, outlining

his opinion that Proctor's merits as an astronomer did not justify his candidacy for the medal.

26. Details of the protracted fight, including the letters between the interlocutors, can be found in vol. 11 of the *Astronomical Register*, 57–67 (quotes on 58 and 66), 93–102, 120–23, 128–30, 181–84. See also Meadows, *Science and Controversy*, 96–103; Hollis, "Decade 1870–1880," 178–84. Hollis's official history is coy about Proctor's removal from the post of honorary secretary. Officially, Proctor resigned by letter while away on a lecture tour of the United States. However, it is likely given the official statement of censure against Proctor's editorial work that he jumped before he was pushed.

27. Pritchard to Lockyer, Nov. 18, 1873, as quoted in Meadows, *Science and Controversy*, 103.

28. Bigg, "Staging the Heavens"; Secord, "Life on the Moon"; Secord, "Planet in Print"; Secord, "Paper Comets."

29. Secord, "Knowledge in Transit." See also Cooter and Pumphrey, "Separate Spheres and Public Places"; Secord, "Electronic Harvest"; Topham, "Scientific Publishing"; Topham, "Rethinking the History"; Topham, introduction to *Isis* Focus section, "Historicizing 'Popular Science'"; Shapin, "Science and the Public"; Fyfe and Lightman, *Science in the Marketplace*; Knight, "Scientists and Their Publics"; Hilgartner, "Dominant View of Popularization"; Lightman, *Victorian Popularizers of Science*; Secord, *Victorian Sensation*; Fyfe, *Science and Salvation*; Morus, "Manufacturing Nature"; Mussell, *Science, Time, and Space*; O'Connor, *Earth on Show*; Keene, "Familiar Science"; Cantor et al., *Science in the Nineteenth-Century Periodical*; Cantor and Shuttleworth, *Science Serialized*; Henson et al., *Culture and Science*; Lewenstein, "From Fax to Facts"; O'Connor, "Reflections on Popular Science"; Taub, *Science Writing in Greco-Roman Antiquity*, 1–21; Frasca-Spada and Jardine, *Books and the Sciences in History*; Hochadel, "Making of a Magic Mountain."

30. On "identity formation" in Victorian science, see Golinski, *Experimental Self*; Barton, "'Men of Science'"; White, *Thomas Huxley*; Ellis, *Masculinity and Science in Britain*; Sponsel, *Darwin's Evolving Identity*; Morus, "Different Experimental Lives"; Aubin and Bigg, "Neither Genius nor Context Incarnate"; Jardine, "Between the Beagle and the Barnacle"; Shapin, "Image of the Man of Science"; Thurs, *Science Talk*; Gieryn, "Boundary Work."

31. Meadows, *Science and Controversy*, is a thorough biography. Recent analysis of Lockyer's idiosyncratic career and his use of diverse public platforms can be found in Bigg, "Staging the Heavens"; Bigg, "Travelling Scientist, Circulating Images"; Gooday, "Sunspots, Weather."

32. The two journals were founded within months of one another. The *English Mechanic* is discussed in greater detail later in this chapter.

33. MacLeod, "Seeds of Competition," 431, 434; Meadows, *Science and*

Controversy, 16–23. After the journal was bought out in late 1864, its list of proprietors included Lockyer, Thomas Huxley, John Tyndall, Francis Galton (a group that J. D. Hooker called the "Young Guard of Science"), plus John Lubbock, Herbert Spencer, William Spottiswoode, and Charles Darwin. On the X Club's self-fashioning through the periodical press, see Kaalund, "Frosty Disagreement."

34. Baldwin, *Making "Nature"* is the definitive account of the journal. See also MacLeod, "Securing the Foundations"; Meadows, *Science and Controversy*, 25–38.

35. Proctor, "Autobiographical Sketch," sent to Charles Kent, Aug. 20, 1873, Letters of Richard Anthony Proctor, PR2—Miscellaneous Papers, Saint John's College Library, Cambridge, UK, hereafter cited as SJCL-RAP; Lightman, *Victorian Popularizers of Science*, 300–307. Proctor's research work has received little analysis. His (controversial) nomination for the RAS Gold Medal is evidence that it was well received by at least some of his peers, as is the fact that, for example, he spoke on "Star-Grouping, Star-Drift, and Star-Mist" for the Royal Institution's prestigious Friday Evening Discourses in Physical Sciences, as reproduced in Lovell, *Royal Institution Library of Science*, 103–12. Proctor's work on the structure of the universe was influential on, for example, Agnes Mary Clerke's *System of the Stars*. See Brück, *Agnes Mary Clerke*, 89–91.

36. Proctor, "Autobiographical Sketch," sent to Charles Kent, Aug. 20, 1873, SJCL-RAP.

37. Lane, "Geographers of Mars"; Lane, "Mapping the Mars Canal Mania"; Lane, *Geographies of Mars*, 26–28.

38. Crowe, *Extraterrestrial Life Debate*. Carver, "Rethinking History," cogently links such analogical reasoning to narratives of the life history of other worlds, a theme picked up later in this chapter and in ch. 4.

39. Lightman, *Victorian Popularizers of Science*, 307–17, quote on 309; Lightman, "Visual Theology of Victorian Popularizers," 665. "Mars, the Miniature of Our Earth" was a chapter title in Proctor, *Other Worlds Than Ours*, and it was a phrase he repeated often in subsequent works. For a globe based on Proctor's 1867 map, see the globe by Busk in figure I.1.

40. Proctor, *Remarks on Browning's Stereograms*, 7–8. Around the same time as this short piece, Proctor articulated his views on the plurality of worlds more generally in two long essays: "Other Habitable Worlds," *Saint Pauls* 3 (Oct. 1868): 47–57; "Other Inhabited Worlds," *Saint Pauls* 3 (Mar. 1869): 676–86.

41. Proctor, *Other Worlds Than Ours*, 45.

42. Proctor, "Autobiographical Sketch," sent to Charles Kent, Aug. 20, 1873, SJCL-RAP.

43. Schaffer, "Where Experiments End," 263–83, quote on 268.

44. Review of *Other Worlds Than Ours*, by Richard Anthony Proctor, *Saturday Review* 29 (Jun. 18, 1870): 806–7, 806. On the conservative and "sometimes slashingly critical" stance of the *Saturday Review*, see Powell, "Saturday Review." For a

similarly critical attack on "sensational science" by the same journal, see the 1875 review cited in Lightman, *Victorian Popularizers of Science*, vii–viii.

45. Dawson and Lightman, *Victorian Scientific Naturalism.*

46. Porter, "Fate of Scientific Naturalism."

47. Baldwin, "Shifting Ground of *Nature*"; Baldwin, "Successors to the X Club"; Baldwin, *Making "Nature,"* 21–47; Bigg, "Travelling Scientist, Circulating Images."

48. Csiszar, "How Lives Became Lists"; Csiszar, *Scientific Journal,* 223–39.

49. Pritchard to Lockyer, May 12, 1870, Sir Norman Lockyer Research Papers, EUL MS 110, Special Collections, University of Exeter Library, Exeter, UK, hereafter cited as UEL-NL.

50. Pritchard, review of *Other Worlds Than Ours,* by Richard Anthony Proctor, *Nature* 2 (Jun. 30, 1870): 161–62.

51. Proctor, letter to the editor, "Other Worlds Than Ours," *Nature* 2 (Jul. 7, 1870): 190.

52. [Lockyer], footnote inserted beneath Proctor's Jul. 7, 1870, letter in *Nature,* 190.

53. Bigg, "Staging the Heavens"; Bigg, "Travelling Scientist." I say "ostensibly" here because, in matter of fact, Alexander Macmillan was remarkably patient in his support of Lockyer's publishing enterprises, which incurred heavy losses for many years (see ch. 1, n59, herein).

54. Proctor, letter to the editor, "Congratulations," *English Mechanic* 11 (Jun. 24, 1870): 323. Proctor does not explicitly name the target of his ire, but there can be little doubt that it was *Nature.* Bernard Lightman contrasts the aims and editorial practices of Lockyer's *Nature* with Proctor's *Knowledge* in Lightman, *Victorian Popularizers of Science,* 325–51.

55. Barton, "Just before Nature," identifies this 1860s boom as "a high point in popular science periodical publishing in nineteenth-century England," a claim that must be treated carefully. Although Barton is right to note the massive expansion of titles in this decade, the majority of the journals she analyses struggled to survive into the 1870s, and none made it to the turn of the century. Only in the 1870s and particularly the early 1880s did the market for popular scientific periodicals stabilize.

56. Proctor, letter to the editor, "Congratulations," 323; review of *Other Worlds Than Ours,* by Richard Anthony Proctor, *English Mechanic* 11 (Jun. 3, 1870): 241. On the *English Mechanic,* see Ewalts, "*English Mechanic*"; Mussell, *Science, Time, and Space,* 29–36. On the importance of correspondence columns to popular science journals in this era, see Lightman, "Popularizers, Participation," 354–55. Lightman, *Victorian Popularizers of Science,* 333n111, notes the likely overlap between the readers of Proctor's books and the *English Mechanic.*

57. Brock, "Science, Technology and Education," 5. Proctor carried this idea to

his own journal, writing an occasional "Paradox Corner" for *Knowledge*. On Proctor's long-winded fight with the flat-earther John Hampden in the *English Mechanic*, see Garwood, *Flat Earth*, 146–50. Parallels might be drawn here with John Pepper's use of displays of public illusions to expose the fraudulence of spiritualists. See Lightman, "Lecturing in the Spatial Economy of Science," 121–24.

58. Clodd, "In Memoriam: Richard Anthony Proctor," *Knowledge* 11 (Oct. 1, 1888), 265; H. G. Salop, letter to the editor, "The Earth's Figure and Rotation," *English Mechanic* 11 (May 20, 1870): 212.

59. Proctor, letter to the editor, "Unfair Criticism," *English Mechanic* 11 (Jul. 8, 1870), 374. In 1868 the *English Mechanic* were boasting a circulation of 80,000, and in 1870 they claimed that their readership was "larger than all other English scientific publications put together." See Brock, "Science, Technology and Education," 3; MacLeod, "Support of Victorian Science," 224. Ewalts, "*English Mechanic*," puts their circulation at a more realistic but still impressive thirty thousand. At the same time, *Nature* was a struggling upstart that was losing money, with perhaps as few as two hundred subscribers and a circulation certainly an order of magnitude smaller than the *English Mechanic*'s. See Meadows, *Science and Controversy*, 25–38; Lockyer and Lockyer, *Life and Work*, 45–50; Baldwin, *Making "Nature,"* 21–73. *Nature* did not turn a profit until 1899, thirty years after it was founded, surviving only because of the largess of the Macmillan publishing house, and as late as 1913 it still had less than a thousand subscribers. See MacLeod, "Securing the Foundations," 443–44; Morgan, *House of Macmillan*, 84–87.

60. Pritchard to Lockyer, Jul. 16, 1870, UEL-NL.

61. Daston, "Objectivity and the Escape"; Daston, "Fear and Loathing"; Daston and Galison, *Objectivity*. I am somewhat skeptical of Daston and Galison's claims regarding imagination in Victorian science. Their study's focus on the atlases at the heart of established, collective, mainstream work risks missing the vibrant and diverse marketplaces for science that many recent studies of Victorian science have explored and articulated in detail. For one cogent criticism, see Willis, *Vision, Science and Literature*.

62. Proctor, *Wages and Wants*, 18, 20; Lightman, *Victorian Popularizers of Science*, 321–24.

63. Herschel, *Treatise on Astronomy*, 1; Herschel, *Outlines of Astronomy*, 2. On Herschel's career and unrivaled public stature, see Cannon, "John Herschel and the Idea of Science"; Chapman, *Victorian Amateur Astronomer*, 53–74. On his complicated relationship with speculation and theory, see Brooke, "Natural Theology," 246; Hoskin, "John Herschel's Cosmology."

64. Crowe, "William and John Herschel's Quest," 257.

65. Proctor, "The Study of Astronomy," *Fraser's Magazine* 4 (Sep. 1871): 282–92, 282 (emphasis mine); "Sir John Herschel," *English Mechanic* 13 (May 19, 1871):

195; "Sir John Herschel as a Theorist in Astronomy," *St. Paul's Magazine* 8 (Jul. 1871): 326–39, 327.

66. Proctor to Herschel, Jul. 24, 1869, as quoted in Proctor, *Other Suns Than Ours*, 401–2.

67. Herschel to Proctor, Aug. 1, 1869, as quoted in Proctor, *Other Suns Than Ours*, 402, 405. Proctor first published excerpts from his long correspondence with Herschel in "Sir John Herschel as a Theorist," 336.

68. Proctor, "The Study of Astronomy," *Fraser's Magazine* 4 (Sep. 1871): 282–92, 282 (emphasis mine).

69. Proctor, "Study of Astronomy," 284.

70. Proctor, *Wages and Wants*, 13, 30, 70–90 (emphasis mine); [Proctor], "Popular Astronomy," *Knowledge* 1 (Feb. 17, 1882): 336–37, 336. See also Lightman, *Victorian Popularizers of Science*, 318–24.

71. White, *Thomas Huxley*, 170–71; Barton, "'Men of Science.'" Proctor's conception of a science worker might be usefully contrasted with Daston and Galison's ideal scientific persona from the latter nineteenth century, the "indefatigable worker": *Objectivity*, 44, 216–33.

72. Both Daston, "Fear and Loathing," and Lindquist, "Visual 'Imagination,'" open their accounts of imagination in Victorian science with Tyndall's 1870 address, which was entitled "Discourse on the Scientific use of the Imagination." Tyndall's cautious approach is noted in Saarloos, "Virtues of Courage," 113–15. See also Flint, *Victorians and the Visual Imagination*, 22–23.

73. Tait's attack was of course made in the pages of *Nature*. As quoted in Rankin and Barton, "Tyndall, Lewes," 51. See also Saarloos, "Virtues of Courage," 118–24; Shapin, "Science and the Public," 996.

74. Tyndall, "Virchow and Evolution," *Nineteenth Century* 4 (Nov. 1878): 809–33, 810. On Tyndall's complex maneuvering within Victorian scientific milieus, see Lightman and Reidy, *Age of Scientific Naturalism*; Dawson and Lightman, *Victorian Scientific Naturalism*; Gieryn, "Boundary Work," 783–87; Turner, "Victorian Conflict between Science and Religion."

75. Proctor, *Poetry of Astronomy*, v.

76. Proctor, *Science Byways*, ix; Proctor, "Study of Astronomy," 286.

77. [Newcomb], review of *The Borderland of Science*, by Richard Anthony Proctor, *Nation* 18 (Mar. 12, 1874): 177. Although published anonymously, Newcomb's considerable body of work for the *Nation* has subsequently been attributed to him by Archibald, "Simon Newcomb, 1835–1909." Newcomb is discussed in greater detail in chs. 2 and 4 herein.

78. [Newcomb], "[Mr. Richard A. Proctor's Lecture Tour]," *Nation* 18 (Apr. 16, 1874): 251–52.

79. Fyfe and Lightman, *Science in the Marketplace*; Morus, "Manufacturing Nature"; Morus, "Seeing and Believing Science"; Morus, "Worlds of Wonder";

Kember, Plunkett, and Sullivan, *Popular Exhibitions*. On visual versus mathematical astronomy and the decline of public interest in the latter, see Donnelly, "Boredom of Science."

80. Proctor, *Science Byways*, xiv (emphasis mine).

81. Schaffer, "Nebular Hypothesis"; Shapin, "Nibbling at the Teats of Science"; Lightman, "Science and the Public," 362–66; MacLeod, "Evolutionism, Internationalism"; Howsam, "Experiment with Science."

82. "The Progress of Physical Science," *Manchester Guardian*, Apr. 4, 1874, 5. The four books reviewed are Proctor, *Moon*; Proctor, *Light Science for Leisure Hours*; Miller, *Romance of Astronomy*; Tyndall, *Six Lectures on Light*.

83. The repeal of the various duties that are colloquially known as the "taxes on knowledge" was finally completed with the removal of the "security system" in 1869. See Hewitt, *Dawn of the Cheap Press*; Lee, *Origins of the Popular Press*, 42–72. The impact of these changes and of liberal education reform are discussed in Curran and Seaton, *Power without Responsibility*, 25–48; Altick, *English Common Reader*, 348–64; Williams, *Long Revolution*, 177–229; Jones, *Powers of the Press*; Brake, Jones, and Madden, *Investigating Victorian Journalism*; Brown, "Treatment of the News"; Brown, *Victorian News and Newspapers*; Secord, "Progress in Print"; Wiener, *Americanization*, 102–28. On technological developments in the British press in this era, see Wiener, "Introduction," xii; Ellis, "Print Paper Pendulum," 3–17; Hobsbawm, *Age of Capital*, 77.

84. Hampton, "Rethinking the 'New Journalism.'" The best arguments for the progressive development of new journalism after 1855 include Wiener, *Papers for the Millions*; Wiener, *Americanization*; Brake, "Old Journalism and the New"; Nicholson, "Provincial Stead." The incorrect dating of new journalism to the 1880s likely derives in large part because the term itself was not coined until 1887, when Matthew Arnold used it in a pejorative sense to critique the journalistic techniques of editors such as William Stead (as discussed later in this chapter). See Baylen, "'New Journalism.'"

85. On these general trends, see Hopkins, *Social History*, 135–49. On the political background and the rise of "popular Liberalism" and socialism in the 1860s, see Hobsbawm, *Age of Capital*, 88–92, 122–42, 245–69; Joyce, *Visions of the People*. On the growth of the Women's Movement in the 1860s, see Jordan, *Women's Movement and Women's Employment*, 145–97. On the expansion after 1850 of mechanics' institutes as sites of working-class nontechnical education, see Laurent, "Science, Society and Politics." On the 1867 Reform act and the 1870 Education Act, see Hoppen, *Mid-Victorian Generation*, 237–71, 591–637. On Mayhew's seminal book, see Thompson, "Political Education of Henry Mayhew." On the *Lancet's* commission of enquiry into London workhouse conditions, see Hodgkinson, *Origins*, 469–82. On the genesis and impact of Smiles's *Self-Help*, see Jarvis, *Samuel Smiles*.

86. Curran, "Capitalism and Control"; Curran and Seaton, *Power without Responsibility*, 32–48; Chalaby, *Invention of Journalism*, 71–193.

87. There is considerable debate over how political the late Victorian press was in comparison to its unstamped forbearers. Chalaby, *Invention of Journalism*, 76–114, on the one hand argues that the British press was largely "depoliticized" after 1855. In contrast, Schudson, "News, Public, Nation"; Schalck, "Fleet Street in the 1880s"; and Curran, "Capitalism and Control," all point to a continued and serious interest in politics by the press after 1855.

88. Curran and Seaton, *Power without Responsibility*, 38–48, quote on 39; Chalaby, "Journalism as an Anglo-American Invention"; Joyce, *Visions of the People*, 65–74.

89. Escott, *Masters of English Journalism*, 251–52. On the trends underlying these shifts in content and ideology, see Koss, *Rise and Fall*, 306–55. On Dickens's populist journalism, see Slater and Drew, "Introduction"; Drew, "Nineteenth-Century Commercial Traveller," 50–61, 83–110. Proctor was a huge fan of Dickens, writing, for example, a book-length tribute to his final novel (Proctor, *Watched by the Dead*). On the *Pall Mall Gazette*'s early campaigns, see Diamond, "Precursor of the New Journalism"; Knelman, *Twisting in the Wind*, 157–80; Greenwood, *Seven Curses of London*, 29–57.

90. As Wiener notes in "How New Was the New Journalism?," 48, it is this commitment to social concerns that above all distinguished the new journalism from other forms of cheap tabloid journalism.

91. Escott, *Masters of English Journalism*, 254. On Stead, see Baylen, "W. T. Stead and the 'New Journalism'"; Baylen, "'New Journalism'"; Baylen, "Stead, William Thomas"; Boston, "W. T. Stead and Democracy"; Dawson, "Stead, William Thomas"; Whyte, *Life of W. T. Stead*; Schults, *Crusader in Babylon*; Eckley, *Maiden Tribute*; Luckhurst et al., *W. T. Stead*. On the rise of the personal in new journalism, in particular through the foregrounding of the editorial persona, see Brake, "Old Journalism and the New"; Salmon, "'A Simulacrum of Power.'" On Morley's surrender of editorial power to Stead as early as 1880, see Robinson, *Muckraker*, 44–47.

92. Stead, "Government by Journalism," *Contemporary Review* 49 (May 1886): 653–74, 663. On Stead's moral crusading, see Baylen, "'New Journalism,'" 382–84, 368–69; Robinson, *Muckraker*. Stead's belief in open access to knowledge included his promotion of indexing systems for the rationalization of knowledge databases and his subsequent projects of knowledge abstraction and synthesis through compilation journals like his *Review of Reviews*. See Dawson, "*Review of Reviews*"; Baylen, "W. T. Stead as Publisher."

93. Mussell, "Stead and the Tabloid Campaign," 24.

94. Otto Struve to Newcomb, Sep. 1, 1881, as quoted in Batten, *Resolute and Undertaking Characters*, 191. Struve wrote in the wake of Christie's appointment

that "the other competitors of whom there was talk (Stone, Proctor, Lockyer) from here indeed seem to be very questionable."

95. Robinson, *Muckraker*, 30–37, 71–73, 121–24, 241–49; Tilley, "Christianity, Journalism, and Popular Print"; Nerone and Barnhurst, "Stead in America." On situating the rise of new journalism within broader cultural and political transformations of the era, see Campbell, "Gladstone, W. T. Stead, Matthew Arnold."

96. Baylen, "'New Journalism,'" 367. See also Robinson, *Muckraker*, 36; Campbell, "Gladstone, W. T. Stead, Matthew Arnold."

97. "Vain Preaching," *Saturday Review* 70 (Nov. 22, 1890): 575–76, 575.

98. Pettitt, *Dr. Livingstone, I Presume?*, 12, 87–91.

99. Wiener, *Americanization*, 28–53, quote on 34; Stevens, *Sensationalism and the New York Press*, 3–53; Schudson, *Discovering the News*, 12–60; Schiller, *Objectivity and the News*; Crouthamel, *Bennett's "New York Herald"*; Guarneri, *Newsprint Metropolis*, 13–53.

100. Chalaby, "Journalism as an Anglo-American Invention"; Riffenburgh, *Myth of the Explorer*, 129; "The Trial for Abduction at the Old Bailey," *Times* (London), Nov. 9, 1885, 9. On the Americanization of the British press after 1870, see Wiener, *Americanization*; Wiener and Hampton, *Anglo-American Media Interactions*. Stead's editorial debt to U.S. journalism is also chronicled in Nerone and Barnhurst, "Stead in America"; Dawson, "*Review of Reviews*"; Gooday, "Profit and Prophecy"; Baylen, "W. T. Stead as Publisher."

101. Bell, "Project for a New Anglo Century"; Bell, "Dreaming the Future"; Bell, *Dreamworlds of Race*, ch. 3. Other advocates of Anglo-American unity included H. G. Wells and Arthur Conan Doyle. Many thanks to Duncan Bell for sharing his insights and forthcoming work on this subject.

102. Stead, *Americanization of the World*, 1–2, 290–93.

103. Saum, "Proctor Interlude in St. Joseph." In the spring of 1880, for example, Proctor gave at least 186 lectures across the entire North American continent. See "Prof. Proctor," *New York Times*, May 25, 1880, 4. Proctor boasted of having visited "five hundred towns" across the United States in his "Froude on America," *New York Tribune*, Aug. 16, 1886, 6. On Proctor's remote editing, see Mussell, "Arthur Cowper Ranyard," 346–47.

104. Proctor, "Capital and Culture in America," *Fortnightly Review*, n.s., 44 (Aug. 1888): 260–78, 272–73. Attendance figures for Proctor's Lowell lectures are given in "Professor Proctor in Boston," *New York Tribune*, Nov. 8, 1875, 1.

105. Proctor to Newcomb, Feb. 2, 1875, container 36, Simon Newcomb Papers, MSS34629, Library of Congress, Washington, D.C., hereafter cited as LOC-SN. Richard Proctor, "The English and American Transit Campaigns Compared," *English Mechanic* 20 (Feb. 12, 1875); Proctor, "True and False Loyalty," *St. Louis Globe-Democrat*, Jul. 24, 1887, 14.

106. Proctor, "Capital and Culture," 272.

107. Scott, "Popular Lecture and the Creation of a Public"; Curti, *Growth of American Thought*, 335–57, quote on 348; Pandora, "Popular Science in National and Transnational Perspective," 350; Rossiter, "Benjamin Silliman and the Lowell Institute." On the broader context for these changes, see Zochert, "Science and the Common Man"; Schiller, *Objectivity and the News*; Kaplan, *Politics and the American Press*; Tucher, *Froth and Scum*. On the social, commercial and religious stratification of the lecture and exhibition scene in Victorian Britain, see Lightman, "Lecturing in the Spatial Economy of Science"; Altick, *Shows of London*, 363–89; Kember, Plunkett, and Sullivan, *Popular Exhibitions*; Morus, Schaffer, and Secord, "Scientific London"; Hays, "London Lecturing Empire"; Secord, *Victorian Sensation*, 439–55; Hewitt, "Beyond Scientific Spectacle." One firm example is the Royal Institution and the "demolished staircase," removing planned access for mechanics and other artisans. See Russell, *Science and Social Change*, 151–52.

108. Channing, *Works of William E. Channing*, 160.

109. "Lectures and Lecturers," *Putnam's Monthly* 9 (Mar. 1857): 317–21, 317, 318. For more on the rise of the lyceum lecture system, see Ray, *Lyceum and Public Culture*; Bode, *American Lyceum*. Lucier, "Professional and the Scientist," 713, notes that lecturers could earn considerable amounts of money from the circuit, often commanding $50 to $150 per lecture and delivering more than a hundred lectures per season. Exemplary of this class of practitioner is Ormsby MacKnight Mitchel, discussed briefly in the next chapter.

110. Advertisement for the Sunday Lecture Society, *Examiner and London Review*, Dec. 31, 1870, 846. On the SLS, see Barton, "Sunday Lecture Societies."

111. "The Sunday Lecture Society," *Examiner and London Review*, Dec. 31, 1870, 841; "The Sunday Lecture Society," *Examiner*, Jan. 6, 1872, 15. On the decline in lecturing in Britain after the 1830s, see Hewitt, "Beyond Scientific Spectacle," 81–82; Lightman, "Lecturing in the Spatial Economy of Science"; Hays, "London Lecturing Empire," 111–12. On Proctor's vocal and disputatious criticism of Sabbatarian Christian orthodoxy, see Bush, "Proctor-Parkes Incident."

112. Kofron, "*Daily News*"; Wiener, "How New Was the New Journalism?," 62; Wiener, *Americanization*, 93–96, 112; Koss, *Fleet Street Radical*, 33–35; Koss, *Rise and Fall*, 192–94, 209–14; Robinson, *Muckraker*, 28–31; Brown, "Treatment of the News."

113. Proctor's regular contributions to the *Daily News* (London), which appeared in the paper anonymously, are noted in, for example, review of *Light Science for Leisure Hours*, by Richard Anthony Proctor, *Manchester Guardian*, Jun. 28, 1871, 7; "Sketch of R. A. Proctor," *Popular Science Monthly* 4 (Feb. 1874): 486–91, 489. A good overview of this work's diversity can be ascertained from the thirteen *Daily News* pieces reproduced in Proctor, *Light Science for Leisure Hours*. Proctor's work for the *New York World* is discussed in the next chapter.

114. Lightman, *Victorian Popularizers of Science*, 325–51, quote on 331.

115. Wiener, "Americanization of the British Press," 66–67; Wiener, *Americanization*, 129–53. Lightman, *Victorian Popularizers of Science*, 348–50.

116. "Are Women Inferior to Men?," *Knowledge* 1 (Nov. 4, 1881): 6–8 and (Nov. 18, 1881): 47–48. Correspondence on the subject can be found on pages 77–78, 95, 165–66 and 456–57 of this volume.

117. Stead's most famous campaign concerning women's rights was the infamous 1885 "Maiden Tribute of Modern Babylon" exposé of widespread child prostitution in London. See Robinson, *Muckraker*, 71–108; Eckley, *Maiden Tribute*, 49–102; Mussell, "Stead and the Tabloid Campaign"; Örnebring, "Maiden Tribute." The "Maiden Tribute" articles were syndicated in the *New York Sun*, typical of the widespread exchange of materials in the transatlantic media marketplace. On Stead and feminism, see Easley, "W. T. Stead, Late Victorian Feminism." On "the sex question" and female suffrage, see Brake, "Writing Women's History." On wider post-*Origin* debates over the scientific evidence for women's inherent inferiority to men, see Conway, "Stereotypes of Femininity"; Richards, "Redrawing the Boundaries."

118. Koss, *Rise and Fall*, 343.

119. "Literary and Art Gossip," *Northern Echo* (Darlington), Jul. 30, 1877, 3; "Easy Star Lessons," *Pall Mall Gazette*, Mar. 11, 1882, 5; "Studies of Venus-Transits," *Pall Mall Gazette*, Aug. 16, 1882, 4–5; "Mr. Proctor's New Books," *Pall Mall Gazette*, Jul. 3, 1883, 4–5. The upturn in positive coverage of Proctor's work after Stead's arrival at the *Gazette* is marked. Before July 1880 the paper carried only a single review of a book by Proctor; in the two years after Stead's arrival, it published at least six reviews, all highly positive.

120. "Current Literature," *Daily News* (London), Apr. 8, 1871, 2.

121. "Studies of Venus-Transits," *Pall Mall Gazette*, Aug. 16, 1882, 4–5; "Mr. Proctor's New Books," *Pall Mall Gazette*, Jul. 3, 1883, 4–5.

122. Proctor was explicit on this point in, for example, his letter to the editor, "Projections v. Formulae," *English Mechanic* 21 (May 14, 1875): 221.

123. Review of *Other Suns Than Ours*, by Richard Anthony Proctor, *Pall Mall Gazette*, Dec. 2, 1887, 3. The Stead quote is from an anonymous entry (and therefore almost certainly by Stead) in the monthly journal he edited after the *Pall Mall Gazette*: "Prussian Annals," *Review of Reviews* 7 (Jan. 1893): 53.

124. "The Magazines of the Month," *Pall Mall Gazette*, Jan. 4, 1884, 11. Proctor's original criticism of Cayley's address can be found in "Dream Space," *Gentleman's Magazine* 256 (Jan. 1884): 35–46.

125. "Mr. Proctor's New Books," *Pall Mall Gazette*, Jul. 3, 1883, 4–5.

126. Wallace to Darwin, May 14, 1871, as quoted in Burkhardt and Secord, *Correspondence of Charles Darwin*, 372–73 (italics in the original).

127. On the target audiences of new journalism, see Jackson, *George Newnes and the New Journalism*, 54–55; Hampton, "Representing the Public Sphere,"

15–29. Mays, "Disease of Reading and Victorian Periodicals," 166, notes that a national debate over "good" and "bad" reading habits peaked around 1885–86. On science in the late Victorian working-class press, see also McLaughlin-Jenkins, "Common Knowledge."

128. Lightman, "Science and the Public," 362–66, quote on 365. On Victorian working-class participation in science, see Secord, "Science in the Pub"; Secord, "Corresponding Interests"; Alberti, "Amateurs and Professionals"; McLaughlin-Jenkins, "Walking the Low Road"; Rose, *Intellectual Life*, 70–72.

129. Crowe, *Extraterrestrial Life Debate*, 373–77; Lightman, *Victorian Popularizers of Science*, 311–13; Carver, "Rethinking History."

130. Proctor is quoted here in a report of his first lecture on the subject in a series of five: "Royal Institution of Great Britain," *Daily News* (London), May 12, 1874, 6. Debates about the age, evolution, habitable period, and eventual "heat death" of Earth were prevalent in this period and fed into Proctor's work. See Brush, "Nebular Hypothesis"; Burchfield, *Lord Kelvin*; Smith and Wise, *Energy and Empire*, 552–645. On the legacy of Proctor's theory and its similarity to Percival Lowell's "planetology," see ch. 4 herein.

131. Proctor, *Life and Death*, 4. I am indebted to Martin Bush for providing me with this source. Bush, "Proctor-Parkes Incident," 31, further discusses Proctor's lectures on cosmic evolution and the life and death of planets.

132. Morus, "Seeing and Believing Science"; Morus, "Philosophy of Demonstration"; Morus, "Worlds of Wonder"; Lightman, "Visual Theology of Victorian Popularizers"; O'Connor, *Earth on Show*; Willis, *Vision, Science and Literature*; Flint, *Victorians and the Visual Imagination*; Smith, *Charles Darwin and Victorian Visual Culture*; Kember, Plunkett, and Sullivan, *Popular Exhibitions*.

133. Nead, *Haunted Gallery*, 199–245, quote on 214.

134. Flint, *Victorians and the Visual Imagination*, 1–39, 62–63; Willis, *Vision, Science and Literature*, 1–9.

135. Proctor, *Poetry of Astronomy*, 291–92. There are strong parallels between this facet of Proctor's imaginative astronomy and the work of his French compatriot Camille Flammarion. As Nead, *Haunted Gallery*, notes on 219 and 233, Flammarion was a master of this kind of remote point of view. Crowe, *Extraterrestrial Life Debate*, 378, calls Flammarion a "French Proctor." However, my own research has not presented Flammarion as a significant figure within British or American astronomical debates during Proctor's lifetime, in large part because it was not until the 1890s that Flammarion's major works, such as his blockbuster *Astronomie Populaire*, were translated into English. Regular columns by Flammarion also began to appear in the English-language Paris edition of the *New York Herald* in this decade and were widely syndicated and extracted across transatlantic media. These publications are discussed in the following two chapters.

136. Jardine, "Made Real." The quotes are from Nasmyth and Carpenter, *The Moon*, plate 2, xiv, 158.

137. Nasim, "James Nasmyth on the Moon"; Robertson, "Science and Fiction"; Boyle, "You Saw the Whole of the Moon"; Carver, "Rethinking History," 443–46; Terpak, "Imaging the Moon." This kind of reorientation would eventually play into and feed off of trends in scientific romances of this era, although this happened for the most part after Proctor's death. See Fayter, "Strange New Worlds"; Markley, *Dying Planet*, 115–49; Crossley, *Imagining Mars*; and the section "New Messages from Mars" in ch. 2 herein.

138. Proctor, *Remarks on Browning's Stereograms*, 3, 6; [Proctor], "The Planet Mars: An Essay by a Whewellite," *Cornhill Magazine* 28 (Jul. 1873): 88–100, 97.

139. See the special issue introduced by Bigg and Vanhoutte, "Spectacular Astronomy." See also Willis, *Staging Science*.

140. The quote is from an advertisement for one of Proctor's lecture tours. See *Knowledge* 8, 340. Proctor earned more than five thousand pounds from a single lecture tour of Australia and New Zealand in May to December 1880. See Bush, "Proctor-Parkes Incident," 27. On the popularity of illustrated astronomical lectures—which often incorporated animated slides that could depict unfolding events—see Nead, *Haunted Gallery*, 223; Willis, "'What the Moon Is Like,'" 179; Hackmann, "Magic Lantern for Scientific Enlightenment"; Hackmann, "Spectacular Science through the Magic Lantern"; Solnit, *River of Shadows*, 200–38; Kember, *Marketing Modernity*, 46–68; Butterworth, "Astronomical Lantern Slides."

141. Hewitt, "Beyond Scientific Spectacle."

142. Kember, *Marketing Modernity*, 46–68.

143. "Best Books for Holiday Reading," *Pall Mall Gazette*, Aug. 2, 1886, 4.

144. "Mr. Proctor's New Books," *Pall Mall Gazette*, Jul. 3, 1883, 4–5. On the International Scientific Series, see Lightman, "Science and the Public," 362–66; MacLeod, "Evolutionism, Internationalism"; Howsam, "Experiment with Science."

145. "Magazines of the Month," 11. On debates in this era over the influence of science journalism on popular readers, see Dawson, "*Review of Reviews*," 184–85.

TWO: Annihilating Time and Space

1. "Richard A. Proctor Has Come to Stay," *New York Tribune*, Jun. 15, 1886, 4.

2. Saum, "Proctor Interlude in St. Joseph," provides useful detail on Proctor's personal life. It appears that his first wife died in January 1879, an event which led him to immediately contemplate moving to Canada or the United States, a move stymied because he "could not readily free [himself] from engagements" in England (see Proctor to Newcomb, Feb. 18, 1879, container 36, LOC-SN). During a convalescence trip to Australia in 1880 Proctor met his future second wife, Sallie Crawley, a native of Saint Joseph. Saum does not mention the Manhattan home, but letters from Proctor to Anna Draper (see ch. 2, n41, herein) were sent from the

Gorham Building in New York City, an up-market apartment house that opened in the spring of 1884.

3. "Meeting of the Royal Astronomical Society, April 14 1882," *Observatory* 5 (May 1, 1882): 125–38, 135.

4. Lane, *Geographies of Mars*, 39–44, 202. See also Tucker, *Nature Exposed*, 210–11; Crowe, *Extraterrestrial Life Debate*, 487–90.

5. Airy, letter to the editor, "The Endowment of Research," *English Mechanic* 32 (Feb. 23, 1881): 586–87, 587.

6. Lockyer, "The Opposition of Mars," *Nature* 46 (Sep. 8, 1892): 443–48.

7. Ball, *In Starry Realms*, as quoted in Aubin, Bigg, and Sibum, *Heavens on Earth*, 1. On Ball as popular author, see also Lightman, *Victorian Popularizers of Science*, 397–421. For Proctor's continued speculations about life on Mars, see, for example his "Life in Mars," *Knowledge* 5 (May 2, 1884): 303–4 and 343–45; "Varied Life in Other Worlds," *Open Court* 1 (Nov. 24, 1887): 595–600.

8. Proctor to Airy, Jun. 26, 1872, as quoted in Meadows, *Science and Controversy*, 98 (see also 112–33); Schaffer, "A World Elsewhere," quote at 25:50; Anderson, *Predicting the Weather*, 235–84, quote on 265. Interest in weather prediction through sunspot cycle analysis was linked to British imperial concerns over catastrophic famines in India. See Davis, *Late Victorian Holocausts*, 216–26; Schaffer, "Where Experiments End," 283–91; Gooday, "Sunspots, Weather." Proctor's criticism of the supposed links between sunspots and terrestrial weather remained vociferous and vocal. See, for example, his "The Weather and the Sun," *Saint Pauls* 13 (Jul. 1873): 99–112; "Sun-Spot, Storm, and Famine," *Gentleman's Magazine* 241 (Dec. 1877): 693–714; "Sun-Spots and Financial Panics," *Scribner's Monthly* 20 (Jun. 1880): 170–78.

9. Higgitt, "British National Observatory"; Hutchins, *British University Observatories*, 219–318.

10. Lankford, *American Astronomy*; Lankford and Slavings, "Industrialization of American Astronomy"; Gingerich, *General History of Astronomy*; Nisbett, "Business Practice"; DeVorkin, *Henry Norris Russell*; "Big Telescope Age"; Sheehan, *Immortal Fire Within*; Osterbrock, *James E. Keeler*; Osterbrock, *Yerkes Observatory*; Osterbrock, Gustafson, and Unruh, *Eye on the Sky*; Wright, *Explorer of the Universe*; Staley, "Michelson and the Observatory"; Bigg, "Spectroscopic Metrologies." On the relative lack of coverage of Mars, see, for example, Lankford's synoptic history of this community, which entirely ignores Mars and the canal debates. One exception is Brush, "Looking Up," which notes in passing on 51 that Schiaparelli's discovery was "made in Italy but exploited in America."

11. Lane, *Geographies of Mars*, 65–95. On the role of philanthropy in the establishment of American observatories, see Miller, *Dollars for Research*, 98–118; Nisbett, "Business Practice," 15–51; Plotkin, "Pickering, the Henry Draper Memorial"; Plotkin, "Pickering and the Endowment of Scientific Research." For a

contemporary opinion on the growing superiority of U.S. observatories over those in European, see Henry Rowland's 1876 letter to James Clerke Maxwell, quoted in Reingold, *Science in Nineteenth Century America*, 269–70.

12. Lane, *Geographies of Mars*, 65–95; Markley, *Dying Planet*, 70–87; Strauss, *Percival Lowell*, 197–203; Willis, *Vision, Science and Literature*, 61–71.

13. Hetherington, "Mid-Nineteenth-Century American Astronomy," 69–76; Warner, "Astronomy in Antebellum America"; Shoemaker, "Stellar Impact"; Goldfarb, "Science and Democracy," 173. Along with Nichol's *Architecture of the Heavens*, Proctor cited Mitchel's *Popular Astronomy* as a formative influence. See the posthumously published "Autobiographical Notes," *New Science Review* 1 (Apr. 1895): 393–97, 393.

14. The Lazzaroni were a loose cohort of like-minded American scientists, centered around Bache (who was the superintendent of the federally funded Coast Survey) and mutually interested in securing control over and federal money for their own elite practices, particularly in relation to geophysics and astronomy. See Bruce, *Launching of Modern American Science*, 217–24; Beach, "Was There a Scientific Lazzaroni?"; Reingold, "Alexander Dallas Bache"; Miller, Voss, and Hussey, *Lazzaroni*; Lurie, *Louis Agassiz*; Jansen, *Alexander Dallas Bache*.

15. Gould, as quoted in Olsen, "Gould Controversy at Dudley Observatory," 269. See also James, *Elites in Conflict*. On Bessel and the European campaign for the primacy of celestial mechanics, see Schaffer, "Where Experiments End," 260–67.

16. Warner, "Astronomy in Antebellum America," 67–68; Rothenberg, "Organization and Control," 309.

17. Reingold, "Definitions and Speculations," 37–38. See also Lucier, "Professional and the Scientist," 700–703, for a useful summary of these debates; and Barton, "'Men of Science,'" which offers support for Reingold's analysis from a British perspective.

18. Highham, "Matrix of Specialization," 16.

19. Reingold, *Science, American Style*, 19–22, 96–126; Menand, *Metaphysical Club*, 57–61. See also Reingold, *Science in Nineteenth Century America*, 59–161. The figure of 144 observatories is given in Clerke, *Popular History of Astronomy*, 8.

20. "Reception to Richard A. Proctor," *New York Times*, Jan. 4, 1874, 5; "Mr. Proctor in Brooklyn," *New York Herald*, Jan. 13, 1874; "Professor Proctor's Closing Lectures," *New York Daily Graphic*, Mar. 17, 1874, 119.

21. Newcomb, *Reminiscences of an Astronomer*, 64. Biographical detail on Newcomb can be found in: Moyer, *Scientist's Voice in American Culture*; Carter and Carter, *Simon Newcomb*; Dick, *Sky and Ocean Joined*, 274–92; Norberg, "Simon Newcomb's Early Astronomical Career"; Norberg, "Simon Newcomb's Role."

22. Newcomb, "The Place of Astronomy among the Sciences," *Sidereal Messenger* 7 (Jan. 1888): 14–20 and (Feb. 1888): 65–72, 69–70. On the prospects of celestial mechanics as a "completed" science, see Badash, "Completeness of

Nineteenth-Century Science"; Schaffer, "Where Experiments End," 260–67. On Newcomb's wider critiques of imaginative astronomy, see, for example, his reviews of Proctor quoted in ch. 1 herein and *"Popular Astronomy* by O. M. Mitchel," *Atlantic Monthly* 6 (Jul. 1860): 117–19; "A Very Popular Astronomer," *Nation* 59 (Dec. 20, 1894): 469–70. On Newcomb as America's preeminent scientist, see J. McKean Cattell, "A Statistical Study of American Men of Science," *Science* 24 (Nov. 23, 1906): 658–65; and (Nov. 30, 1906): 699–707; and (Dec. 7, 1906): 732–42, which collates results of a survey sent to American scientists asking them to grade their peers, and places Newcomb unanimously first among astronomers and first overall for all American scientists. Cattell's lists are anonymous, but notes made by Newcomb's family, held in LOC-SN, container 125, identify "Astronomer I" as Newcomb.

23. "Recent Literature: Proctor's Universe, Proctor's Expanse of the Heavens," *Atlantic Monthly* 34 (Sep. 1874): 363–64, 363 (emphasis mine); Proctor, letter to the editor, "Mr. R. A. Proctor and the Atlantic Monthly," *New York Times*, Dec. 31, 1874, 4. See also Proctor, "To the Editor of the Atlantic Monthly," *Atlantic Monthly* 34 (Dec. 1874): 750–51. Proctor used his favorite British outlet to carry on the counterattack on both sides of the Atlantic. See his letter to the editor, "Saturn," *English Mechanic* 21 (Apr. 2, 1875), 62. On the *Atlantic Monthly*'s high-minded intellectual ideals, see Sedgwick, "Atlantic Monthly." On support for Proctor during the furor, see, for example, "Professor Proctor," *Chicago Daily Inter Ocean*, Feb. 9, 1875, 4; "Proctor and the Atlantic Monthly," *Hartford Daily Courant*, Jan. 1, 1875, 1; "Richard A. Proctor and the Atlantic Monthly," *Boston Daily Advertiser*, Jan. 2, 1875; "Occasional Notes," *Pall Mall Gazette,* Jan. 25, 1875, 5; "Premature Criticism," *Popular Science Monthly* 7 (May 1875): 124.

24. Proctor to Newcomb, Jun. 7, 1875, container 36, LOC-SN (emphasis mine).

25. Proctor to Newcomb, Jan. 5, 1876, container 36, LOC-SN.

26. Proctor to Newcomb, Jan. 6, 1876, container 36, LOC-SN (italics in the original).

27. Proctor to Newcomb, Jan. 6, 10, 12, 1876, container 36, LOC-SN.

28. Jones and Boyd, *Harvard College Observatory*, 305; Osterbrock, "Rise and Fall of Edward S. Holden," 83–84; Gingerich, "Satellites of Mars," 113. Proctor later made this embarrassing episode public in a letter to the editor, "Note from Mr. Proctor," *Sidereal Messenger* 6 (Sep. 1887): 259–62, 260.

29. Osterbrock, "Rise and Fall of Edward S. Holden," 83–86.

30. Staley, "Michelson and the Observatory," 223–34. Rowland's 1883 talk before the American Association for the Advancement of Science, which was pointedly entitled "A Plea for Pure Research," is quoted on 233–34.

31. Thurs, *Science Talk*, 1–21. See also Gieryn, "Boundary Work."

32. [Mitchel], "Introductory Remarks," *Sidereal Messenger* 1 (Jul. 1846): 1–2, 1; "Address of Benjamin Apthorp Gould, Ex-President of the Association,"

Proceedings of the American Association for the Advancement of Science 18 (1870): 1–37, 18.

33. McCormmach, "Ormsby MacKnight Mitchel's 'Sidereal Messenger'"; Curti, *Growth of American Thought*, 348.

34. On the development and growth of the U.S. telegraph network, see Thompson, *Wiring a Continent*; Standage, *Victorian Internet*. On the telegraph's impact on society, see Kern, *Culture of Time and Space*; Mussell, *Science, Time and Space*; Solnit, *River of Shadows*; Schaffer, "Time Machines"; Carey, *Communication as Culture*, 155–77; Marvin, *When Old Technologies Were New*; Otis, *Networking*; Otis, "Metaphoric Circuit."

35. Stachurski, *Longitude by Wire*; Bartky, *Selling the True Time*, 32–89; Stephens, "Most Reliable Time"; Stephens, "Astronomy as Public Utility"; Reingold, *Science, American Style*, 113; Loomis, *Recent Progress of Astronomy*, 304–67.

36. Jones and Boyd, *Harvard College Observatory*, 194–98. In Europe the central distribution point was the Royal Observatory in Kiel, Germany.

37. Newcomb to Lockyer, Nov. 18, 1877, UEL-NL.

38. Guarneri, *Newsprint Metropolis*, 13–53, 194–233; Blondheim, *News over the Wires*; Schwarzlose, *Nation's Newsbrokers*; Carey, *Communication as Culture*, 162–63. Global perspectives can be found in Winseck and Pike, *Communication and Empire*; Hobsbawm, *Age of Capital*, 75–77; Putnis, Kaul, and Wilke, *International Communication*.

39. Strauss, *Percival Lowell*, 234–35; R. M. McCreary, "Newspaper Astronomy," *Popular Astronomy* 6 (Apr. 1898): 103–4, 103. Numerous examples of newspaper coverage of astronomical events can be found in Cottam and Orchiston, *Eclipses, Transits, and Comets of the Nineteenth Century*.

40. Nisbett, "Business Practice," 15–28; Jones and Boyd, *Harvard College Observatory*, 211–45; Boyd, "Mrs. Henry Draper"; Plotkin, "Pickering, the Henry Draper Memorial"; Becker, *Unravelling Starlight*, 179–82, 193, 224–25.

41. Proctor to Anna Draper, Dec. 21, 23, 1884, Henry and Mary Anna Palmer Draper Papers, MssCol 838, New York Public Library, New York, hereafter cited as NYPL-HMAPD. In New York, Proctor had a particularly close relationship with the *Tribune* and the *World*. See Saum, "Proctor Interlude in St. Joseph," 41, and the discussion of Proctor's work for the *World* later in this chapter. When at his residence in Saint Joseph, Proctor was a regular contributor to the *St. Louis Globe-Democrat*, writing at least sixty-eight articles for the paper between September 1886 and December 1887.

42. Wright, *James Lick's Monument*, 28, 117–20. See also Osterbrock, Gustafson, and Unruh, *Eye on the Sky*; Sheehan, *Immortal Fire Within*, 110–40.

43. Stevens, *Sensationalism and the New York Press*, 57–100; Schudson, *Discovering the News*, 88–106; Juergens, *Joseph Pulitzer*; Fettmann, *"New York World"*; Guarneri, *Newsprint Metropolis*, 106–13; Churchill, *Park Row*, 39; Erickson,

"Yellow Journalism," 607. After buying the paper in 1883, Pulitzer rapidly overhauled the *World,* increasing its circulation to more than a quarter of a million within three years and a million within a decade.

44. The precursor to all these failures was, of course, the most famous giant telescope ever constructed, the "Leviathan of Parsonstown," built by the Third Earl of Rosse in the early 1840s and fitted with a six-foot mirror. After the impressive discovery of spiral nebulae only months into its life, the Leviathan subsequently endured a disappointing career characterized by a terrible site, mirror-tarnishing problems, and its unwieldy design, all faults pointed out by Proctor himself. See Schaffer, "On Astronomical Drawing," 467. See also Schaffer, "Leviathan of Parsonstown"; King, *History of the Telescope,* 206–17; Hoskin, "Rosse, Robinson," 339; Hoskin, "First Drawing of a Spiral Nebula"; Bennett, "Era of Newton, Herschel and Lord Rosse," 41; Chapman, *Victorian Amateur Astronomer,* 96–100. On Lassell's forty-eight-inch reflector in Malta, which was dismantled after only three years, see King, *History of the Telescope,* 220–24; Chapman, "William Lassell." In 1877 Lassell wrote to Henry Draper that he feared the astronomical telescope had already reached its maximum size. See Wright, *James Lick's Monument,* 44–45. Cooke's twenty-five-inch Newall refractor failed at a poor site in Newcastle. See Van Helden, "Telescope Building, 1850–1900," 46–47; King, *History of the Telescope,* 251–54; Lequeux, "Great Nineteenth Century Refractors," 50. Grubb's twenty-four-inch reflector for Edinburgh observatory suffered through a series of design compromises. See Glass, *Victorian Telescope Makers,* 69–71. On the Toulouse and Paris failures, see Lequeux, 58–59. The trials and tribulations of the Great Melbourne Telescope are chronicled in Gillespie, *Great Melbourne Telescope*; Glass, *Victorian Telescope Makers,* 39–61; Hyde, "Calamity of the Great Melbourne Telescope," quote on 229; Gascoigne, "Great Melbourne Telescope"; Schaffer, "Easily Cracked," 714–15; Royal Society, *Correspondence concerning the Great Melbourne Telescope*; King, 264–67. On Huggins's work on nebulae, see Becker, *Unravelling Starlight,* 64–81. Proctor was also likely aware of Henry Bessemer's disastrous attempts to revolutionize large mirror production, including his destruction of a fifty-inch mirror by Calver. See Van Helden, 57; H. P. Hollis, "Large Telescopes," *Observatory* 37 (1914): 245–52; "Three Giant Telescopes," *Times* (London), Apr. 3, 1880, 6.

45. Proctor, "Proctor on the Big Lens: He Thinks the Lick Telescope Will Disappoint Science," *New York World,* Feb. 27, 1887, 17.

46. Procter, *William Randolph Hearst,* 40–58. In its first year under Hearst the paper doubled its circulation (53). By 1893 the *Examiner* was outselling its largest city rival, the *Chronicle,* to become the fourth largest paper in the country. See also Riffenburgh, *Myth of the Explorer,* 96; Collins, *Murder of the Century,* 31, 144.

47. "The Lick Telescope," *San Francisco Daily Examiner,* Feb. 28, 1887, 2; "The Big Telescope," *San Francisco Daily Examiner,* Mar. 1, 1887, 5. The first of

these two pieces carried the dateline "New York, February 26" and was marked "Special to the *Examiner*."

48. "A Scorcher: Professor Proctor Replies to Unfair Strictures," *San Francisco Daily Examiner*, Mar. 27, 1887, 4. Proctor himself publicized the clash with Holden through his own journal in England, his regular column in the *St. Louis Globe-Democrat*, as well as anywhere where these pieces were picked up and excepted. See Proctor, "The Great Lick Telescope," *Knowledge* 10 (Jul. 1887): 205–7; "Gossip," *Knowledge* 10 (Jul. 1887): 209–10; Proctor, "Large Telescopes," *St. Louis Globe-Democrat*, Sep. 25, 1887, 22.

49. "Prosperity in California," *New York Daily Tribune*, Mar. 6, 1887, 12; editorial, "Professor Proctor," *New York Daily Tribune*, Mar. 13, 1887, 4; [Payne], "The Holden–Proctor Unpleasantness," *Sidereal Messenger* 6 (May 1887): 192; Holden, letter to the editor, "President Holden's Reply to Professor Proctor," *Sidereal Messenger* 6 (Jun. 1887): 210–12. (Payne's journal is not to be confused with O. M. Mitchel's journal of the same title from the 1840s.)

50. Proctor, letter to the editor, "Note from Mr. Proctor."

51. Jones and Boyd, *Harvard College Observatory*, 246–68; Osterbrock, *Yerkes Observatory*, 8–9; Nisbett Becker, "Professionals on the Peak," 493–99. Boyden's endowment did not specify where the observatory should be sited—only that it should be at altitude. This bequest and its consequences are discussed in detail in the next chapter. Spence's plan eventually collapsed after the 1891 Los Angeles property crash, but not before a world-beating forty-inch lens had been ordered. This lens was eventually purchased by the railroad tycoon Charles Yerkes on behalf of the University of Chicago (as discussed later in this chapter).

52. The full title of Solnit's book is *River of Shadows: Eadweard Muybridge and the Technological Wild West*. The second quote is on 123.

53. Adas, *Dominance by design*, 67–127, is the key study of technology's implication in westward imperial expansion. Vetter, *Field Life*, is the best study of scientific research and western conquest. On the great power of western industrial capitalism, see Robbins, *Colony and Empire*, 83–102. Slotkin, *Fatal Environment*, cogently analyses the tension between the "frontier myth" and American industrial conquest of the West and its native residents. Cronon, *Nature's Metropolis*, is the definitive study of the role of expanding metropolitan commerce in the transformation of the "Great West." Solnit, *River of Shadows*, 3–24, 66, 123, details the links between robber-baron philanthropy and technological glorification of the West. On "manliness" and western frontier heroism, see Lears, *Rebirth of a Nation*, 12–45.

54. Quoted in Wright, *James Lick's Monument*, 29.

55. *San Francisco Chronicle*, Apr. 14, 1886, as quoted in Wright, *James Lick's Monument*, 118. On the "annihilation of time and space" as a trope of the technological West, see Solnit, *River of Shadows*, 3–24, 179–205; Belknap, *From a*

Photograph, 121–65. On railways and the "Industrialization of Time and Space," see Schivelbusch, *Railway Journey*.

56. Lane, "Astronomers at Altitude"; Lane, *Geographies of Mars*, 65–95. As Holden's *Mountain Observatories* recounts, there had been since midcentury a series of *temporary* high-altitude observation sites established in Europe and North America, of varying degrees of success.

57. Turner, *Significance of the Frontier*, 1–5.

58. "A Dignified Reply," *San Francisco Daily Examiner*, Mar. 28, 1887, 4; "The Retort Courteous: Professor Holden Replies to the Letter of Mr. Proctor," *San Francisco Daily Examiner*, Mar. 28, 1887, 4; "Holden and Proctor," *San Francisco Daily Examiner*, Mar. 1, 1887, 2 (emphasis mine).

59. Wright, *James Lick's Monument*, 10–11, 25. On Davidson, see Lewis, *George Davidson*; Smith, *Pacific Visions*. On the characteristics of this "distinctive western science," see White, *New History of the American West*, 128; and Vetter, *Field Life*.

60. Lane, *Geographies of Mars*, 65–95; Lears, *Rebirth of a Nation*, 31–45; Martin, "Buffalo Bill's Wild West"; Solnit, *River of Shadows*, 155–76; Rydell and Kroes, *Buffalo Bill in Bologna*, 105–11.

61. Solnit, *River of Shadows*, 24.

62. "Holden and Proctor," *San Francisco Daily Examiner*, Mar. 1, 1887, 2. The *Daily Alta* is quoted in Wright, *James Lick's Monument*, 11.

63. Osterbrock, *Yerkes Observatory*, 6; Cronon, *Nature's Metropolis*, xvi–xvii. It is notable that although the Yerkes Observatory was sited well away from Chicago to escape the poor seeing conditions inherent to any built-up area, the site chosen by University of Chicago President William Rainey Harper was not at altitude. Harper's choice at the edge of Geneva Lake was evidently influenced by factors other than seeing—including the proximity to the summer homes of wealthy potential benefactors. Despite receiving assurances from certain astronomers that the lake's environment would not be a problem, this proved to be false, and the site was never ideal for such a large instrument. See Osterbrock, *Yerkes Observatory*, 15–16.

64. Clark, "Great Telescopes of the Future," *Astronomy and Astro-physics* 12 (Oct. 1893): 673–78, 673; Osterbrock, *Yerkes Observatory*, 11–15. On Clark see Warner, *Alvan Clark and Sons*.

65. Rydell, *All the World's a Fair*, 38–71; Rydell and Kroes, *Buffalo Bill in Bologna*, 47–72; Greenhalgh, *Ephemeral Vistas*.

66. Proctor, "Varied Life in Other Worlds," *Open Court* 1 (Nov. 24, 1887): 595–600. Quoted is the journal's strap line. On *Open Court*'s quest for a "religion of the future" through ecumenical synthesis, see Lears, *Rebirth of a Nation*, 238. Proctor left his intended magnum opus, *Old and New Astronomy*, unfinished at the time of his unexpected death in September 1888. It was eventually completed by Proctor's friend and colleague, A. Cowper Ranyard, and published posthumously in 1892.

67. "The New Born Parliament," *New York Times*, Aug. 7, 1892, 1.

68. Solnit, *River of Shadows*, 19. Lankford and Slavings, "Industrialization of American Astronomy," and Nisbett, "Business Practice," both suggest that another significant feature of this transformation of new astronomy was the incorporation into astrophysics of "business practice" models of organization and management, both for the handling of large staffs and the processing of vast quantities of data.

69. Willis, *Vision, Science and Literature*, 5.

70. Flammarion, "Mars and Its Inhabitants," *North American Review* 162 (1896): 546–57, 557. On Martian signaling, see also Crowe, *Extraterrestrial Life Debate*, 393–400; Lane, *Geographies of Mars*, 197–201; Crossley, *Imagining Mars*, 58–65; Schroeder, "Message from Mars," 24–29.

71. Flammarion, "Inter-Astral Communication," *New Review* 6 (1892): 106–14, 106. Flammarion also promoted the prize in "Idee d'une communication entre les mondes," *Astronomie* 10 (1891): 282–87; "Invisible Worlds," *New York Herald*, Jun. 4, 1891, 6; "Long Distance Signals," *New York Herald*, Jul. 12, 1891, 13; "How to Talk with the Folks on Mars," *New York Herald*, Jan. 3, 1892, 7; "Possibilities of Planetary Inhabitants," *New York Herald*, May 29, 1892, 30.

72. Galton, letter to the editor, "Sun Signals to Mars," *Times* (London), Aug. 6, 1892, 7. As Tattersdill, *Science, Fiction*, notes, Galton had experimented with mirror signaling while traveling in Africa in his youth (30). A typical newspaper response to Galton's letter is "Earth May Signal Mars," *New York Times*, Aug. 7, 1892, 5. A London correspondent suggested periodic dimming of that city's lights as the simplest means of flashing a signal. See H. R. Haweis, letter to the editor, "How to Speak with Mars," *Pall Mall Gazette*, Aug. 18, 1892, 2.

73. Flammarion, "Inter-Astral Communication," 107–8. On the late nineteenth-century trope of the superior Martian "casting a penetrating reverse gaze towards the Earth," see Lane, *Geographies of Mars*, 193–97.

74. Flammarion, "Inter-Astral Communication," 112–14. On the new technologies of long-distance communication, see Marvin, *When Old Technologies Were New*, 184–90; Sconce, *Haunted Media*, 95–103. Spiritualist communication with Martians would become a hot topic. See, for example, "News from Mars: Alleged Communications by a Martian 'Control,'" *Borderland* 4 (1897): 406–9; Flournoy, *From India to the Planet Mars*; Crossley, "Mars and the Paranormal," 466–84; Tattersdill, *Science, Fiction*, 49–53.

75. "Reading the Stars a la Mode," *Punch* 103 (Aug. 20, 1892): 78. Lane, *Geographies of Mars*, 199 discusses the range of cultural products that engaged with interplanetary communication in this period. Tesla discussed the practicalities of signaling Mars with an enthusiastic press in 1899, then reported receiving messages from another planet a year later. See Carlson, *Tesla*, 264–65, 274–78, 315; Tesla, "Talking with the Planets," *Collier's Weekly*, Feb. 9, 1901, 4–5.

76. "Notes: The Opposition of Mars," *Journal of the British Astronomical Association* 2 (1892): 477.

77. Lockyer, "Opposition of Mars," 443 (emphasis mine).

78. "A Strange Light on Mars," *Nature* 50 (Aug. 2, 1894): 319. The French telegram reads, "Light projection in southern region of the terminator of Mars observed by Javelle 28 July 4 am Perrotin" (my translation).

79. Galton, "Intelligible Signals between Neighbouring Stars," *Fortnightly Review* 60 (Nov. 1896): 657–64, 657–59 (italics in the original).

80. Tattersdill, *Science, Fiction*, 40–41, notes that Galton's initial draft lecture on interplanetary signaling lacked this newspaper framework, which was then added when the author rewrote the piece for the *Fortnightly Review*, "re-framing it for a large audience in the language of a mass media format which cast its readers as witnesses to developing current events." On seriality in nineteenth-century science, see the special issue introduced by Hopwood, Schaffer, and Secord, "Seriality and Scientific Objects."

81. Galton, "Intelligible Signals Between Neighbouring Stars," 657; W. L. Courtney to Galton, Oct. 2, 1896, Papers and Correspondence of Sir Francis Galton, GB 0103 GALTON, University College Library, Special Collections, London, UK, hereafter cited as UCL-FG, 241/9. Galton's reply does not survive, but in a follow-up letter from Courtney, dated Oct. 12, 1896, the editor readily acquiesces to the author's preference: "By all means let your article stand as it is at present. Mine was only a suggestion for you to consider."

82. Willis, *Visions, Science and Literature*, 57–113.

83. Wells, "Popularising Science," *Nature* 50 (Jul. 26, 1894): 300–301.

84. Wells to Allen, [late summer 1895?], as quoted in Smith, *Correspondence of H. G. Wells*, 1:245–46. There are a number of in-depth studies that consider the links between Wells's science journalism and his science fiction, including McLean, *Early Fiction*; Willis, *Mesmerists, Monsters, and Machines*; Willis, *Vision, Science and Literature*, 57–113; Markley, *Dying Planet*; Crossley, "H. G. Wells, Visionary Telescopes"; Crossley, *Imagining Mars*; Haynes, *Wells, Discoverer of the Future*; Tattersdill, *Science, Fiction*. A useful overview of the vast literature on Wells, including a bibliography, can be found in Crossley, "Grandeur of H. G. Wells."

85. [Wells], "Intelligence on Mars," *Saturday Review*, Apr. 4, 1896, 345–46; Wells, "The Crystal Egg," *The New Review* 16 (May 1897): 556–71, 571.

86. Beck, *War of the Worlds*; McLean, *Early Fiction*, 89–113; Willis, *Vision, Science and Literature*, 57–113; Markley, *Dying Planet*, 122–27; Crossley, "H. G. Wells, Visionary Telescopes"; Crossley, *Imagining Mars*, 110–28; Tattersdill, *Science, Fiction*, 48–49.

87. [Wells], "Intelligence on Mars," 345; Wells, *War of the Worlds*, 6. Two notable exceptions to the lack of attention paid to the links between Martian signaling and Wells's novel are Worth, "Imperial Transmissions," and Tattersdill, *Science, Fiction*, 48.

88. Wells, *War of the Worlds*, 18, 21–22, 174; Worth, "Imperial Transmissions," 69–76.

89. "Dwellers in the Planet Mars? Facts, Guesses, and Fiction," *Review of Reviews* 16 (1897): 489. On spiritualist communications with Mars, see ch. 2, n74, herein.

THREE: Constructing Canals on Mars

A version of this chapter was published previously as Nall, "Constructing Canals on Mars."

1. [Clerke], "New Views about Mars," *Edinburgh Review* 184 (1896): 368–85. Attributed to Clerke by the Wellesley Index.

2. Secord, "Knowledge in Transit." See also Secord, *Victorian Sensation*; Secord, *Visions of Science*; Fyfe and Lightman, *Science in the Marketplace*; Fyfe, *Science and Salvation*; Morus, "Manufacturing Nature"; Topham, "Scientific Publishing"; Topham, "Rethinking the History"; Shapin, "Science and the Public"; Lightman, *Victorian Popularizers of Science*; Mussell, *Science, Time, and Space*; O'Connor, *Earth on Show*.

3. Lenoir, *Inscribing Science*; Carey, *Communication as Culture*, 155–77; Poovey, "Limits of the Universal Knowledge Project"; Lewenstein, "From Fax to Facts"; Derrida, "Archive Fever," 17–18; Mussell, *Science, Time, and Space*; Kern, *Culture of Time and Space*.

4. Pang, "Social Event of the Season," 268; Lane, *Geographies of Mars*, 100. See also Lane, "Astronomers at Altitude"; Riffenburgh, *Myth of the Explorer*; Pang, *Empire and the Sun*; Ratcliff, *Transit of Venus Enterprise*.

5. Crouthamel, *Bennett's "New York Herald"*; Blondheim, *News over the Wires*; O'Connor, *Scandalous Mr. Bennett*; Guarneri, *Newsprint Metropolis*, 106–13; Winseck and Pike, *Communication and Empire*, 43–91; Schwarzlose, *Nation's Newsbrokers*.

6. Stanley, *How I Found Livingstone*, xviii; Riffenburgh, *Myth of the Explorer*, 49–137.

7. Jones and Boyd, *Harvard College Observatory*, 176–444; Nisbett, "Business Practice"; Sobel, *Glass Universe*, 3–55.

8. Nisbett, "Business Practice," 52–95, quote on 52; Nisbett Becker, "Professionals on the Peak"; Jones and Boyd, *Harvard College Observatory*, 246–68; Plotkin, "Boyden Station in Peru."

9. "Stars' Strange Secrets Shown to Astronomers," *New York Herald*, Oct. 19, 1889, 5. Other examples of this wide press coverage include "Prof. W. H. Pickering," *New York Evening Post*, Jan. 26, 1889, 9; "Photographing the Southern Heavens," *New York Sun*, Mar. 30, 1890, 15; "Mapping the Southern Sky from a Mountain Peak 14,000 Feet High," *Scientific American* 64 (1891): 36.

10. Edward Pickering was a famously prodigious fundraiser, who went so far

as to hire a public relations expert, George Michaelis, to manage his observatory's newspaper coverage. See Lankford, *American Astronomy*, 189; Plotkin, "Endowment of Scientific Research."

11. Good biographical detail on William Pickering's idiosyncratic career can be found in Jones and Boyd, *Harvard College Observatory*; Plotkin, "William H. Pickering in Jamaica."

12. William Pickering, "Harvard's Success at Willow," *New York Herald*, Jan. 2, 1889, 3; "How the News Was Gathered," *New York Herald*, Jan. 2, 1889, 3; "The Hidden Sun," *New York Herald*, Jan. 3, 1889, 3.

13. "How the News Was Gathered."

14. "Delighted Scientists: The Herald's Enterprise and Its Results Commended," *New York Herald*, Jan. 3, 1889, 3.

15. Todd, "How Man's Messenger Outran the Moon," *Century* 38 (1889): 602–6; reprinted as "The Moon Distanced," *New York Herald*, Aug. 4, 1889, 22.

16. O'Connor, *Scandalous Mr. Bennett*, 184. On post-1880 U.S. imperial intervention in South America, see LaFeber, *American Foreign Relations*, 60–82; Lens, *Forging of the American Empire*, 159–62.

17. Winseck and Pike, *Communication and Empire*, 77–80; Ahvenainen, *European Cable Companies in South America*, 96–114; Berthold, *Telephone and Telegraph in Chile*, 30–33; *Via Galveston*; "Cable Laying Record Broken," *New York Herald*, May 30, 1893, 10.

18. "The Central and South American Telegraph and the Herald's News," *New York Herald*, May 30, 1893, 8; Crouthamel, *Bennett's "New York Herald,"* 48–50, 56–68, 138–41.

19. "The Herald's Eclipse News on Two Continents," *New York Herald*, Apr. 18, 1893, 8; "Results of the Sun's Eclipse," *New York Herald*, Apr. 18, 1893, 9. The *Herald* published at least twelve pieces on the eclipse between February 16 and June 9.

20. William Pickering to Wolfe, Feb. 28, Mar. 6, Oct. 13 (quotation), 1893, UAV-630.14.5-E4, Harvard College Observatory, Records of Director Edward C. Pickering, HOLLIS No. 001966647, Harvard University Archives, Cambridge, MA, hereafter cited as HUA-ECP. See also "Professor Pickering Thanks the Herald," *New York Herald*, Jun. 9, 1893, 8.

21. "To Map Out the Starry Heavens," *New York Herald*, Dec. 15, 1890, 4; "Harvard's Astronomical Expedition," *New York Herald*, Dec. 15, 1890, 6.

22. Jones and Boyd, *Harvard College Observatory*, 246–68, 287–324; Nisbett, "Business Practice," 37–38, 52–150; Plotkin, "Boyden Station in Peru"; Nisbett Becker, "Professionals on the Peak"; Lane, *Geographies of Mars*, 107–15; Fernie, *Whisper and the Vision*, 173–79. Nisbett notes that by the end of his first year in Peru, William had spent twenty-two thousand dollars and had budgeted twelve thousand more for the coming year (108).

23. Sobel, *Glass Universe*.

24. William Pickering, "The Physical Aspect of the Planet Mars," *Science* 12 (1888): 82–84; William Pickering, "Visual Observations of the Surface of Mars," *Sidereal Messenger* 9 (1890): 369–70.

25. Crowe, *Extraterrestrial Life Debate*, 367–86, quote on 386; Markley, *Dying Planet*, 59–60; Bensaude-Vincent, "Camille Flammarion."

26. "Peculiar Events in Mars," *New York Herald*, Sep. 15, 1890, 7; "News from the Planet Mars," *New York Herald*, Sep. 16, 1890, 6; "Other Worlds Than Ours," *New York Herald*, Oct. 6, 1890, 7; "Strange Phenomena Observed on Mars," *New York Herald*, Oct. 7, 1890, 6.

27. "What Professor Holden Says," *New York Herald*, Oct. 9, 1890, 7. See also Holden, "Notes on the Opposition of Mars, 1890," *Publications of the Astronomical Society of the Pacific* 2 (1890): 299–300.

28. "The Planet Mars Is Nigh to Death," *New York Herald*, Oct. 9, 1890, 7. Because the two papers had a syndicate agreement, much of this article also appeared on the same day in Holden's local paper, the *San Francisco Chronicle*, 1. On the trope of Mars as a dying planet, see Markley, *Dying Planet*.

29. "Harvard's Astronomical Expedition," *New York Herald*, Dec. 15, 1890, 6.

30. William Pickering to Edward Pickering, Jun. 1, 1891, UAV-630.14.5-E2, HUA-ECP. See also William Pickering, "Astronomical Possibilities at Considerable Altitudes," *Astronomische Nachrichten* 129 (1892): 98–99.

31. William Pickering, "Investigations in Astronomical Photography," *Annals of Harvard College Observatory* 32 (1895): 1–115, 109; Jones and Boyd, *Harvard College Observatory*, 304. On wider struggles to photograph Mars in this era, see Tucker, *Nature Exposed*, 215–17.

32. William Pickering to Charlotte Pickering, Jul. 17 and 31, 1892, HUG-1691.4, Papers of William H. Pickering, HOLLIS No. 000604354, Harvard University Archives, Cambridge, MA, hereafter cited as HUA-WHP.

33. "Invisible Worlds," *New York Herald*, Jun. 4, 1891, 6; "Long Distance Signals," *New York Herald*, Jul. 12, 1891, 13; "How to Talk with the Folks on Mars," *New York Herald*, Jan. 3, 1892, 7; "Possibilities of Planetary Inhabitants," *New York Herald*, May 29, 1892, 30.

34. Edward Pickering to William Pickering, May 29, 1892, as quoted in Jones and Boyd, *Harvard College Observatory*, 309.

35. This point was often stressed in pieces the *Herald* ran that thanked the paper for its assistance in transmitting William's observations. See "Appreciation of the Herald's Enterprise," *New York Herald*, Aug. 12, 1892, 6; "Herald Service Appreciated," *New York Herald*, Sep. 2, 1892, 7; "Professor Pickering Thanks the Herald," *New York Herald*, Jun. 9, 1893, 8.

36. Nisbett Becker, "Professionals on the Peak."

37. William Pickering to Edward Pickering, Jul. 15, 1892, HUG-1691.4.5,

HUA-WHP. Edward's stern letter of reply wasn't sent until August 7 so would not have reached William until well after the opposition of Mars was already over.

38. "Observations of Mars from Greenwich," *Baltimore Sun*, Aug. 12, 1892, 2; "Mars and His Satellites," *Baltimore Sun*, Aug. 2, 1892, 1. At U.S. and European latitudes Mars was low in the sky, a point Holden was particularly keen to stress. See "Opposition of Mars," *San Francisco Chronicle*, Aug. 4, 1892, 3.

39. "Watching All Over the Globe," *New York Times*, Aug. 5, 1892, 1. See also "No Canals on the Planet: So Say Observers at the Naval and Lick Observatories," *Washington Post*, Aug. 4, 1892, 2.

40. "Observations in Peru: Harvard's South American Annex Expects to Do Good Work," *New York Herald*, Aug. 3, 1892, 3.

41. "There Is a Bad Side: Professor Holden Points Out the Danger of Expecting Too Much," *New York Herald*, Aug. 3, 1892, 3.

42. "No Great Discoveries in Regards to Mars," *New York Herald*, Aug. 5, 1892, 4.

43. William Pickering, telegram to *New York Herald* office in Lima, Aug. 8, 1892; William Pickering to Baker, Aug. 8, 1892; William Pickering to Wolfe, Aug. 8, 1892, UAV-630.14.5-E3, HUA-ECP; William Pickering to Charlotte Pickering, Aug. 7, 1892, HUG-1691.4, HUA-WHP.

44. "Observations of Mars in South America," *New York Herald*, Aug. 10, 1892, 3; "Large Areas of Blue: Seas Seen One-Half Size of Mediterranean," *Boston Daily Globe*, Aug. 10, 1892, 4; "Many Changes Noted in Fiery Mars," *Chicago Tribune*, Aug. 10, 1892, 1; "Studies of Mars," *St. Louis Post-Dispatch*, Aug. 10, 1892, 2; "As Viewed from Peru: Phenomena Seen in the Red Planet," *San Francisco Chronicle*, Aug. 10, 1892, 1.

45. "Observations of Mars in South America"; "Melting of Mars' Snow-Cap," *New York Herald*, Aug. 10, 1892, 6.

46. "Making New Map: Why Astronomers Watch Planet Mars," *Boston Daily*, Aug. 4, 1892, 4.

47. "Heartless Prof. Holden," *Washington Post*, Aug. 6, 1892, 4 (emphasis mine).

48. "The Folks on Mars," *New York Times*, Aug. 7, 1892, 4.

49. "Hot Shot for Holden: A Brooklyn Paper's Sharp Criticism," *San Francisco Chronicle*, Aug. 11, 1892, 1.

50. Holden, "The Lowell Observatory in Arizona," *Publications of the Astronomical Society of the Pacific* 6 (1894): 160–69, 166.

51. Contrast, for example, the *New York Times* London correspondent's pro-Lick cable of August 7, which noted how "universally precedence has been given the news from the Lick Observatory over all others, and how all take it for granted that the most valuable results of observation come from there" (see ch. 2, n67, herein), with "Color Changes upon Mars: What Prof. Pickering Has Observer While in Peru," *New York Times*, Aug. 30, 1892, 8.

52. "At Lick Observatory: Mapping the Canals of Mars," *Los Angeles Times*, Aug. 19, 1892, 2. Telegraphically networked syndication arrangements ensured that all Associated Press and *Herald* pieces during the great Mars boom appeared simultaneously in most major cities in the United States and Europe. On the role of syndication in nationalizing news gathered by U.S. metropolitan papers, see Guarneri, *Newsprint Metropolis*, 204–9.

53. For example Holden, "Notes on the Opposition of Mars, 1890," *Publications of the Astronomical Society of the Pacific* 2 (1890): 299–300; "A Square Look at Mars: The Lick Telescope Dispels Many Erroneous Theories," *New York Times*, Aug. 1, 1892, 5; "No Canals on the Planet: So Say Observers at the Naval and Lick Observatories," *Washington Post*, Aug. 4, 1892, 2.

54. For example Paul Gibier, "Possible Uses of Mars' Canals," *New York Herald*, Aug. 21, 1892, 22, which speculated that the canals might be an artificial irrigation network, or highways. This report was widely reprinted and excerpted, first through syndication and then through the usual practice of flagrant copying or re-wording by nonsyndicated papers in subsequent days and weeks. Schiaparelli himself, as was typical, remained playfully ambiguous about the implications of the doubling that he saw: "Life on Mars: Prof. Schiaparelli Says 'Tis a Theory," *Boston Daily Globe*, Aug. 5, 1892, 1.

55. "The Double 'Canals' of Mars," *New York Herald*, Aug. 20, 1892, 6.

56. "Melting Snow in Mars: Professor Pickering . . . Makes Many Interesting Discoveries," *New York Herald*, Sep. 1, 1892, 7. William Pickering wrote Holden on September 1 that "we are so out of the world that we haven't heard a word yet of what has been done on Mars by any other Observatory" (UAV-630.14.5-E3, HUA-ECP).

57. "Peeps at a Planet," *Auburn Bulletin* (New York), Sep. 19, 1892, 6.

58. *New York Herald*, Sep. 15, 1892, 8; "Does Reduction Reduce?," *Havana Journal* (New York), Oct. 15, 1892, 4.

59. Holden, "Note on the Mount Hamilton Observations of Mars, June—August, 1892," *Astronomy and Astro-physics* 11 (1892): 663–68, footnote on 663.

60. William Pickering to Edward Pickering, Jul. 7, 1892, HUG-1691.4.5, HUA-WHP.

61. Holden often made this point to the press. See, for example "Opposition of Mars," *San Francisco Chronicle*, Aug. 4, 1892, 3.

62. Carey, *Communication as Culture*, 162–63.

63. Markley, *Dying Planet*, 55.

64. William Pickering, "Visual Observations of the Surface of Mars," *Sidereal Messenger* 9 (1890): 369–70, 369.

65. Holden, for example, argued that if one analyzed drawings of Mars alone, ignoring written descriptions, it became evident that "there are enormous difficulties in the way of completely explaining the recorded phenomena by terrestrial

analogies unless we also introduce serious modifications." See Holden, "Note on the Mount Hamilton Observations of Mars, June–August, 1892," *Astronomy and Astro-physics* 11 (1892): 663–68, 668. On Mars and terrestrial analogy, see Markley, *Dying Planet*, 31–114; Lane, *Geographies of Mars*, 141–85.

66. Lockyer, "Opposition of Mars," 447.

67. Lockyer's argument was soon taken up by other critics of claims about the nature and extent of Mars's canal network, most notably Simon Newcomb and Alfred Russel Wallace. The former made the claim in his articles on Mars for both the tenth and eleventh editions of the *Encyclopaedia Britannica* (as discussed in the next chapter), as well as in various popular books and articles. Wallace made the claim in both his *Wonderful Century* (239) and *Is Mars Habitable?* (3). Recourse to the mistranslation claim is nearly ubiquitous in secondary literature on the canal controversy. I have found at least twenty-five pieces that present it as *the* explanation for the controversy, none of which mention its historical origins in the debate itself. Two recent works that express admirable caution with respect to the claim are Markley, *Dying Planet* (55), and Lane, *Geographies of Mars*, the second chapter of which persuasively posits a visual rather than textual explanation for the interpretation of Schiaparelli's canals as artificial.

68. William Pickering to Edward Pickering, Aug. 8, 1892, HUG-1691.4.5, HUA-WHP. The reference to three hundred thousand dollars here indicates that William was well aware of his brother's recent attempts to start fundraising for a major new telescope for the Arequipa site. See Williamina Fleming, "Harvard College Observatory Astronomical Expedition to Peru," *Publications of the Astronomical Society of the Pacific* 4 (1892): 58–62, 61; *New York Sun*, Sep. 18, 1892, 4; *New York Tribune*, Sep. 19, 1892, 3, and Sep. 20, 6; *New York Times*, Sep. 19, 1892, 9; *New York Herald*, Sep. 26, 1892, 4, 6, and Oct. 15, 10, and Oct. 24, 6, and Dec. 5, 6; Edward Pickering, "A Large Southern Telescope," *Science* 20 (1892): 193–94, also reprinted in *Publications of the Astronomical Society of the Pacific* 4 (1892): 214–17.

69. Edward Pickering to William Pickering, Aug. 24, 1892, as quoted in Jones and Boyd, *Harvard College Observatory*, 307.

70. "Etc.," *Overland Monthly* 20 (1892): 328. Shinn's journal was an important booster for the technological West. Founded in the summer of 1868, its mission was to represent the finest attributes of nature, culture and industry in California. See Clements, "Overland Monthly."

71. More than one hundred astronomical works published in the wake of the 1892 opposition are analyzed in Crowe, *Extraterrestrial Life Debate*, 496–502. Terby is quoted on 499.

72. Lowell is quoted and paraphrased here by Holden in his "The Lowell Observatory in Arizona," *Publications of the Astronomical Society of the Pacific* 6 (1894): 160–69, 161–62.

FOUR: Made to Last

1. Edward Hale, "Latest News from Mars," *Boston Commonwealth*, as reprinted in *Scientific American* 72 (Mar. 2, 1895): 137. Hale's report was widely quoted and reprinted.

2. Strauss, "Founding of the Lowell Observatory"; Dick, *Biological Universe*, 70–75; Putnam, *Explorers of Mars Hill*, 13–42. William Pickering and A. E. Douglass were both given a leave of absence from Harvard Observatory to work with Lowell. Strauss shows that initial plans had envisaged the Flagstaff site as a satellite of Harvard Observatory, an idea Lowell rejected in order to secure control over the site and its output.

3. Chandler to Barnard, Sep. 4, 1894, as quoted in Sheehan, *Immortal Fire Within*, 241; Campbell, "An Explanation of the Bright Projections Observed on the Terminator of Mars," *Publications of the Astronomical Society of the Pacific* 6 (Mar. 31, 1894): 103–12, 103.

4. Lowell, address to the Boston Scientific Society, May 22, 1894, reprinted in the *Boston Commonwealth* and quoted in Sheehan, *Planet Mars*, 106.

5. Holden, "The Lowell Observatory in Arizona," *Publications of the Astronomical Society of the Pacific* 6 (Jun. 9, 1894): 160–69, 160.

6. "The Lowell Observatory," *Observatory* 17 (Sep. 1894): 311–12, 312; "Notes," *Observatory* 18 (Apr. 1895): 177.

7. The first Lowell quote is from a 1905 paper submitted to the Royal Astronomical Society, as quoted in Strauss, "'Fireflies Flashing in Unison,'" 157. The second is from Lowell, *Mars and Its Canals*, 111. On the history of planetology, see also Strauss, *Percival Lowell*, 210–19; Lane, *Geographies of Mars*, 141–85. Crossley, *Imagining Mars*, 70, reports that Lowell's archive contains annotated copies of Proctor's *Other Worlds Than Ours* and Flammarion's *La Planète Mars*. Although historians have barely drawn the link, Lowell's planetology was clearly influenced by Proctor's earlier works on planetary life cycles. Lowell did cite Proctor's studies of Mars from time to time, and he appears to have stolen Proctor's memorable description of Mars "as the abode of life" for the title of his final book on the planet (see Proctor, *Other Worlds Than Ours*, 84). However, Lowell's debt to Proctor has gone almost completely unnoticed, and Proctor's work in this area is largely ignored by accounts of the rise of planetary science, such as Jaki, *Planets and Planetarians;* Brush, *History of Modern Planetary Physics;* and Doel, *Solar System Astronomy*.

8. Markley, *Dying Planet*, 61–114. In public and in private, Lowell frequently equated himself with Darwin and his critics with Owen and other anti-evolution dogmatists.

9. See the section "New Messages from Mars" in ch. 2 herein and Markley, *Dying Planet*, 115–49; Crossley, *Imagining Mars;* Lane, *Geographies of Mars*, 187–216.

10. Lowell as popularizer is a pervasive trope in the literature, and examples

are too extensive to list. Significant examples include Hoyt, *Lowell and Mars;* Hetherington, "Amateurs versus Professionals"; Hetherington, "Percival Lowell"; Hetherington, *Science and Objectivity,* 49–64; Sheehan, *Planet Mars,* 98–113; Crossley, *Imagining Mars,* 127; Lane, *Geographies of Mars,* 7–12.

11. Lowell was elected a member of the Astronomical and Astrophysical Society of America in the first round of elections in 1900. He presented at the society's first conversazione in 1901 (as discussed briefly later in this chapter), gave two papers at the society's third meeting in 1902, two at their sixth meeting in 1904, and one at their eighth meeting in 1906. See *Publications of the Astronomical and Astrophysical Society of America* 1 (1910): xii, 149, 163, 219–21, 283. Alongside a busy schedule of public lectures, Lowell also traveled to present and debate with his astronomical colleagues, speaking before both the RAS and the British Astronomical Association during European tours in 1905 and 1910. See Tucker, *Nature Exposed,* 219, and the meeting reports in *Observatory* 33 (May 1910): 191–204. On his professorship, see Strauss, *Percival Lowell,* 56–57. Lowell angrily corrected Simon Newcomb when the latter referred to him as "Mr." See Lowell to Newcomb, Oct. 3, 1905, container 30, LOC-SN.

12. An important exception to this is DeVorkin, "W. W. Campbell's Spectroscopic Study."

13. Newcomb, "Fallacies about Mars," *Harper's Weekly* 52 (Jul. 25, 1908): 11–12. Newcomb's depiction here is perhaps too black and white. It is worth remembering that some astronomers, such as W. W. Campbell, vehemently opposed Lowell's methodology and "proofs" of Martian canals yet were more than willing to countenance the existence of some form of life on the planet.

14. Moyer, *Scientist's Voice in American Culture.* Newcomb's 1908 article in *Harper's Weekly* carried a subhead that read "Professor Newcomb is, perhaps, the foremost living astronomer. His achievements have won him distinction throughout the world." A flavor of Newcomb's catholic interests can be got from a very comprehensive bibliography compiled in Archibald, "Simon Newcomb, 1835–1909."

15. Adams, *Education of Henry Adams,* 377; Wells, *Time Machine,* 4–5; Fayter, "Strange New Worlds," 264.

16. Adams, *Education of Henry Adams,* 381–82. "The Dynamo and the Virgin," ch. 25 of Adams's autobiography, which recounts his disorienting and revelatory experience of the exposition's Palace of Electricity, is a powerful evocation of the remarkable transformations then underway in the physical sciences of the fin de siècle. For more on these transformations, see Morus, *When Physics Became King,* 156–225; Nye, *Before Big Science,* 147–88; Galison, *Einstein's Clocks, Poincaré's Maps;* Schaffer, "Time Machines"; Kern, *Culture of Time and Space;* Marvin, *When Old Technologies Were New;* Levin et al., *Urban Modernity.* On the Paris Universal Exposition, see Romein, *Watershed of Two Eras,* 296–308; Geppert, *Fleeting Cities,* 62–100; Mandell, *Paris, 1900;* Bennett et al., *1900.* On science and

technology at the great international exhibitions of the era, see Brain, *Going to the Fair*; Brain, "Going to the Exhibition"; Staley, *Einstein's Generation*, 137–65; Nye, "Electrifying Expositions"; Greenhalgh, *Ephemeral Vistas*, 142–73; Rydell, *All the World's a Fair*.

17. Newcomb to Kelvin, Oct. 14, 1907, container 53, LOC-SN. On Kelvin, radiation, and debates over the age of the sun and Earth, see Burchfield, *Lord Kelvin*, 163–211; Smith and Wise, *Energy and Empire*, 549–51, 607–11. On Newcomb's positivist empiricism, see Moyer, *Scientist's Voice in American Culture*, 9–15, 36–51, 82–97.

18. Moyer, *Scientist's Voice in American Culture*, 9–15, 90–97, 188–94, quote on 190.

19. Newcomb to Irving Fisher, Nov. 2 and 10, 1904; Mar. 8, 1907, containers 51 and 53, LOC-SN; Moyer, *Scientist's Voice in American Culture*, 90–94, quote on 191. Newcomb's aspirations to make economics a positive science are discussed in Barber, "Should the American Economic Association." Newcomb consistently stressed the potential for "the opinions of university men . . . in influencing public opinion in the right direction." Newcomb to Charles Eliot, Nov. 18, 1902, container 48, LOC-SN.

20. Newcomb, *His Wisdom the Defender*, 14.

21. See the section of the bibliography herein: "Simon Newcomb's Encyclopedia Work." Newcomb also wrote numerous definitions for at least two dictionaries. See Archibald, "Simon Newcomb, 1835–1909," 65, 67.

22. Solnit, *River of Shadows*, 232.

23. Bigg, "Staging the Heavens."

24. On this point it is worth remembering that most books, even textbooks, rarely went through print runs of more than a few thousand. See Fyfe, "Information Revolution." The eleventh-edition *Encyclopaedia Britannica* (1910–11), in comparison, sold at least three hundred thousand copies. See Walsh, *Anglo-American General Encyclopedias*, 50. By the mid-1890s, the *New York World* had a combined morning and evening circulation of more than a million. See Fettmann, "*New York World*."

25. Geppert, *Fleeting Cities*, quote on 204; Greenhalgh, *Ephemeral Vistas*, 20; Bellon, "Science at the Crystal Focus." On fragmentary knowledge and attempts to order it in this period, see Yeo, "Reading Encyclopedias"; Yeo, "Encyclopaedic Knowledge"; Yeo, *Encyclopaedic Visions*; Secord, *Victorian Sensation*, 41–76; Secord, "Progress in Print"; Secord, "Science, Technology and Mathematics"; Csiszar, "Seriality and the Search for Order"; Mussell, *Science, Time, and Space*, 91–120. On the rise of "middlebrow" consumer culture, see Rubin, *Making of Middlebrow Culture*; Radway, *Feeling for Books*, 127–301; Levine, *Highbrow / Lowbrow*; McKitterick, *History of Cambridge University Press*, 192–93.

26. Proctor, letter to the editor, "Note from Mr. Proctor," *Sidereal Messenger*

6 (Sep. 1887): 259–62. This was certainly a plum job to poach—the *New American Cyclopaedia* (1858–63) and its sequel the *American Cyclopedia* (1873–76) were conceived as a New World response to the major European encyclopedias. See Walsh, *Anglo-American General Encyclopedias*, 2–3, 109–10.

27. Yeo, "Reading Encyclopedias," 43–47.

28. Newcomb, "How an Encyclopaedia May Be Edited," *Nation* 87 (Nov. 19, 1908): 492. Confusing matters was the possibility that this work was a knockoff version of an older encyclopedia, *The American Educator* (1897), for which Newcomb said that he "quite possibly" wrote "one or two articles." I have found three reprints of this 1897 work, under three different titles, published in three different cities, in three different years (see the "Simon Newcomb's Encyclopedia Work" section of the bibliography), highlighting the buoyant U.S. market for knockoff encyclopedias in the gilded age and progressive era. Newcomb wrote for at least one of them, the Werner Company's 1897 unauthorized reprint (with American supplements) of the *Britannica* ninth edition. He also later served as an expert witness for the *Britannica* against Werner (see containers 17, 57, and 58 in LOC-SN). On U.S. piracy of the *Britannica*, see Walsh, *Anglo-American General Encyclopedias*, 52–54; Kogan, *Great EB*, 64–68; Kruse, "Story of the *Encyclopaedia Britannica*," 195–225.

29. Jardine, "Books, Texts, and the Making of Knowledge"; Secord, "Science, Technology and Mathematics"; Frasca-Spada and Jardine, *Books and the Sciences in History*; Topham, "Book History and the Sciences."

30. Yeo, "Reading Encyclopedias," 42; Yeo, *Encyclopaedic Visions*, 246–83. Yeo's focus on Enlightenment encyclopedias is mirrored by Darnton, *Business of Enlightenment*; Collison, *Encyclopaedias*; Hughes, "Science in English Encyclopaedias."

31. McKitterick, *History of Cambridge University Press*, 183–201; Kogan, *Great EB*, 87–207; Walsh, *Anglo-American General Encyclopedias*, 49–50; Kruse, "Story of the *Encyclopaedia Britannica*," 276–324. The work's Americanization was noted in, for example, George Burr's review in *American Historical Review* 17 (Oct. 1911): 103–9, 104.

32. Chisholm to Newcomb, Jan. 4, 1903, container 19, LOC-SN; Einbinder, *Myth of the Britannica*, 48–49. The eleventh edition became "a byword for learned authority" according to McKitterick, *History of Cambridge University Press*, 183.

33. Brain, *Going to the Fair*, 17.

34. Bennett et al., *1900*, 3; Brain, "Going to the Exhibition," 123–27; Rydell and Kroes, *Buffalo Bill in Bologna*, 98–105; Greenhalgh, *Ephemeral Vistas*, 18–22. On the cut-and-paste transatlantic periodicals of the era, such as W. T. Stead's *Review of Reviews* and George Newnes's *Strand Magazine*, see Dawson, "New Journalism"; Baylen, "W. T. Stead as Publisher"; Mussell, *Science, Time and Space*, 61–87; Jackson, *George Newnes*.

35. Loveland, "Unifying Knowledge and Dividing Disciplines," traces the

development in the treatise format over the early edition of the *Britannica*, but he misses the significant shift in presentation of science content in the eleventh, as does Kruse, "Story of the *Encyclopaedia Britannica*." Practicalities of typesetting and printing eventually lead to the eleventh's issue in two tranches in late 1910 and early 1911. See McKitterick, *History of Cambridge University Press*, 190.

36. Yeo, "Reading Encyclopedias," 29–34; Hughes, "Science in English Encyclopaedias," 350. The "tenth edition" of the *Britannica* was not itself a true edition, being merely a reissue of the ninth edition with eleven supplementary volumes tacked on. I briefly discuss the production of these hastily compiled supplements later in this chapter.

37. George Burr's review in *American Historical Review* 17 (Oct. 1911): 103–9, 103.

38. For details of the planning stages of the astronomy content for the eleventh, see LOC-SN, containers 17, 58, and 59. Chisholm agreed with Newcomb's assessment of Proctor's material, and he outlined to Newcomb his ambitions for the science content under its new organizational plan in letters of Mar. 7, Apr. 3, 12, 1905, container 17, LOC-SN.

39. Chisholm, "Editorial Introduction," *Encyclopaedia Britannica*, 11th ed., vol. 1, *A to Androphagi* (Cambridge: Cambridge University Press, 1910), xviii.

40. Newcomb, "Astronomy," *Encyclopaedia Britannica*, 10th ed., vol. 1, *Aachen—Australia* (Edinburgh: Adam and Charles Black, 1902), 728–56, 729–30. Barnard's report on Mars was buried within his "Micrometrical Measures of the Ball and Ring System of the Planet Saturn . . . With Some Remarks on Large and Small Telescopes," *Monthly Notices of the Royal Astronomical Society* 56 (Jan. 1896): 163–72.

41. Newcomb, *Astronomy for Everybody*, 185.

42. Lowell to Newcomb, Mar. 1, 15, 1903, container 30, LOC-SN (emphasis mine); Newcomb to Lowell, Mar. 9, 1903, Early Correspondence of the Lowell Observatory, 1894–1916, Microfilm Edition, Lowell Observatory Archive, Flagstaff, AZ, hereafter cited as LOA-ECLO. For more on Lowell's idea of "acute" versus "sensitive" eyesight, see ch. 4, n49, and the "Astronomy and Visual Representation in the Early Twentieth Century" section later in this chapter.

43. Strauss, *Percival Lowell*, 220–40; DeVorkin, "W. W. Campbell's Spectroscopic Study"; Sheehan, *Immortal Fire Within*, 237–63. Holden faded as an opponent of Lowell after he was forced from the directorship of the Lick in 1897 following much controversy. He was replaced by Campbell and never secured another astronomical post. See Osterbrock, "Rise and Fall of Edward S. Holden." He did keep up the fight against Lowell, however, through such pieces as "What We Know about Mars," *McClure's Magazine* 16 (Mar. 1901): 439–44. William Pickering, meanwhile, after leaving the Lowell Observatory under strained circumstances also wrote in opposition to Lowell's artificial canals. See the various papers reproduced in his career-spanning *Mars*, in particular 103–72. However, his eccentricities and

peripatetic career marginalized him from the astronomical mainstream after 1900. See Plotkin, "William H. Pickering in Jamaica."

44. Hale to Newcomb, Nov. 12, 1901, container 25, LOC-SN. Newcomb actually offered Lowell a place on the organizing committee, as well as exhibition space for his drawings and magic lantern slides of Mars. See Lowell to Newcomb, Nov. 7, Dec. 5, 16, 20, 1901, container 30, LOC-SN. On the foundation of the AAS, see DeVorkin, *American Astronomical Society's First Century*, 3–57; Rothenberg, "Organization and Control."

45. Newcomb to Lowell, Mar. 23, 1903, LOA-ECLO.

46. Proctor, "On Self-Deception in Observation," *English Mechanic* 15 (Jul. 19, 1872): 447–48. Several examples of astronomers raising this issue in the mid-1880s are given in Lankford, "Amateurs versus Professionals."

47. Maunder, "The Canals of Mars," *Knowledge* 17 (Nov. 1894): 249–52, 252. This point had been made earlier by, for example, Charles Young in "Small Telescopes vs. Large," *Sidereal Messenger* 5 (Jan. 1886): 1–5, 2. See also Maunder, "The 'Eye' of Mars," *Knowledge* 18 (Mar. 1895): 55–59; B. W. Lane, "The Canals of Mars," *Knowledge* 25 (Nov. 1902): 250–52, which includes a comment by Maunder. For more on this debate, see also Crowe, *Extraterrestrial Life Debate*, 505–6, 524–26; Crowe, "Astronomy and Religion," 220–24; Ruiz-Castell, "Priority Claims and Public Disputes"; Sheehan, *Planets and Perception*, 224–56.

48. Webb, *Tree Rings and Telescopes*, 44–49.

49. Willis, *Vision, Science and Literature*, 71–83. The expertise quote is from Lowell, *Mars*, 159. Lowell expanded on the issue of "acute" versus "sensitive" eyesight in his "On the Kind of Eye Needed for the Detection of Planetary Detail," *Popular Astronomy* 13 (Feb. 1905): 92–94; and Lowell to Newcomb, Feb. 16, 1905, container 59, LOC-SN.

50. J. E. Evans and Maunder, "Experiments as to the Actuality of the 'Canals' Observed on Mars," *Monthly Notices of the Royal Astronomical Society* 63 (Jun. 1903): 488–99.

51. Key examples include Hoyt, *Lowell and Mars*, 168; Burnett, "British Studies of Mars"; Hetherington, *Science and Objectivity*, 49–64; Sheehan, *Planets and Perception*, 124–256; Sheehan, *Immortal Fire Within*, 253.

52. Flammarion, "Experiences Contre la Realite des Canaux de Mars," *Bulletin de la Societe Astronomique de France* 19 (1905): 274–83, editorial note on 283. Other examples of criticisms of Maunder's paper by allies of Lowell are given in Strauss, *Percival Lowell*, 208.

53. "Professor Hall at Work," *New York Herald*, Aug. 5, 1892, 4. As Markley, *Dying Planet*, 73, and Willis, *Vision, Science and Literature*, 71–83, both note, Lowell and his staff's superior experience as Martian observers was continually stressed by Lowell and his allies, for example in Lowell, *Mars and Its Canals*, 174–77. Nasim, *Observing by Hand*, analyzes the "processes of familiarization" necessary for

drawing enigmatic celestial objects like nebulae. Schaffer, "On Astronomical Drawing," likewise considers the problem of calibrating different observers' sketches made at the eyepiece. Flint, *Victorians and the Visual Imagination*, 29–30, notes that the individual subjectivity of vision and the concomitant importance of training the eye was a much recognized problem in this era.

54. "Report of the Meeting of the Association Held on December 30th, 1903," *Journal of the British Astronomical Association* 14, no. 3 (1904): 111–18, 118. Lowell made the same point more colloquially in *Mars and Its Canals*, 202–3. Lowell declined an invitation to further debate the matter in the *Illustrated London News*. See Tucker, *Nature Exposed*, 215. On the Lowell Observatory vision experiments, see Strauss, *Percival Lowell*, 193–94, 198–201, 224; Webb, *Tree Rings and Telescopes*, 47–48; Hoyt, *Lowell and Mars*, 164–68; Sheehan, *Immortal Fire Within*, 257. The *Lowell Observatory Bulletin* 2 (1903) reports in detail the results of their experiments "On the Visibility of Fine Lines."

55. "The 'Canals' of Mars," *English Mechanic* 77 (Jun. 19, 1903), 407.

56. On the exposition, see Rydell, *All the World's a Fair*, 154–83, quote on 159. On the Congress and American scholarship in the public sphere, see Coats, "American Scholarship Comes of Age." On Newcomb's vision for the Saint Louis Congress, see his "The Coming International Congress of Arts and Sciences at St. Louis," *Popular Science Monthly* 65 (Sep. 1904): 466–73. Newcomb visited Europe from May to September 1903, with ninety-six foreign speakers eventually appearing at Saint Louis (Coats, "American Scholarship Comes of Age," 411). Lowell appears on a list of invitees drawn up by Hale, but it is not clear whether he attended or not. See Hale to Newcomb, Aug. 25, 1904, container 25, LOC-SN.

57. George Edwin Rines to Newcomb, Oct. 4, 1904, container 38, LOC-SN. On the symbiosis of cultures of print and exposition, see part 2 of Kember, Plunkett, and Sullivan, *Popular Exhibitions*.

58. Rines to Newcomb, Jan. 19, 1904, container 38, LOC-SN (italics in the original); Newcomb to Rines, Feb. 11, 1904, container 50, LOC-SN.

59. Nead, *Haunted Gallery*, 219–30; Bigg, "Staging the Heavens"; Solnit, *River of Shadows*, 155–259; Solomon, "Trip to the Fair"; Byrn, *Progress of Invention*, 288; Boyd, *Paris Exposition of 1900*, 551.

60. [Newcomb], "Wallace on Life in the Universe," *Nation* 78 (Jan. 14, 1904): 34–35, 34 (emphasis mine). Similar views are also expressed by Newcomb in his "Life in the Universe," *Harper's Magazine* 111 (Aug. 1905): 404–8.

61. Chisholm to Newcomb, Feb. 9, 1905, container 17, LOC-SN.

62. Newcomb to Lowell, May 8, 1905. See also his letters of Feb. 20 and Apr. 25, and Lowell's responses to Newcomb on Feb. 16 and May 2, containers 30 and 52, LOC-SN.

63. Winter, "Early Spaceflight Simulation Shows," 1:153–54; "Photograph Mars Canals," *New York Times*, May 28, 1905, 1.

64. *New York Times* to Lowell, Apr. 9, 24, 1905, LOA-ECLO. See also Jul. 8, 1907, which references an agreement "on same terms as two years ago," although there is no record of what exactly these terms were.

65. "The Dispute about Mars," *New York Times*, Sep. 30, 1907, 6.

66. Tucker, *Nature Exposed*, 207–33, quote on 207. The tiny, grainy Lowell Observatory photographs of Mars were so hard to reproduce that they were habitually both retouched and presented alongside drawings. This important point is missed by accounts that attempt to directly link the demise of Lowell's theory with the development of "objective" astrophotography, such as Galison, "Judgement against Objectivity," 329–31; Daston and Galison, *Objectivity*, 179–82. Similar problems of intervention and aesthetics pervaded attempts to disseminate astronomical photographs in this era. See Pang, "Victorian Observing Practices," pt. 1 and 2; Pang, "Astrophotography at the Early Lick Observatory"; Pang, "Development of Astrophotography at the Lick"; Mussell, "Arthur Cowper Ranyard"; Belknap, *From a Photograph*.

67. Wesley, "Photographs of Mars," *Observatory* 28 (Aug. 1905): 314–15; [Turner], "From an Oxford Note-book," *Observatory* 28 (Aug. 1905), 332–36, 336; "Photographs of Mars: Negatives Taken in Arizona Clearly Show Canals," *Washington Post*, Jul. 16, 1905, 8; "The Canals in Mars," *Times* (London), Sep. 29, 1905, 2 (also widely reprinted in the United States, including in the *New York Times* and the *Scientific American Supplement*); Hale to Barnard, Nov. 9, 1905, as quoted in Strauss, *Percival Lowell*, 230. See also Markley, *Dying Planet*, 87–92; Crowe, *Extraterrestrial Life Debate*, 527–28. William Pickering in his *Photographic Atlas of the Moon*, 23, noted the dangers inherent to the photographic process, not least that "slight changes in exposure and development will sometimes produce results that are very misleading."

68. Newcomb to Chisholm, Feb. 24, 1905; Clerke to Newcomb, Aug. 2, 1905; containers 19 and 52, LOC-SN. Clerke was commissioned to write the history of astronomy and much of the biographical material for the eleventh edition.

69. Newcomb to Chisholm, Oct. 9, 1905, container 52, LOC-SN.

70. Newcomb to Lowell, Oct. 20, 1905, container 52, LOC-SN. See also Newcomb to Lowell, Sep. 21, 1905; Lowell to Newcomb, Oct. 3, 18, 1905, containers 30 and 52, LOC-SN.

71. Lowell to Newcomb, Oct. 27, 1905 (emphasis mine); Newcomb to Lowell, Oct. 30, 1905, containers 30 and 52, LOC-SN.

72. Lowell to Newcomb, Nov. 6, 1905, container 30, LOC-SN. Lowell suggested two people that Newcomb would never have seriously considered: Lowell Observatory's own "Mr. Lampland, the taker of the photographs of the canals," and the young Scottish journalist Hector MacPherson, who had written glowingly of Lowell and his artificial canal hypothesis. Lowell's confidence in the photographs

coincides with his *Bulletin* on the subject being reprinted in a major public forum. See "The Canals of Mars—Photographed," *Popular Astronomy* 13 (Nov. 1905): 479–84. This was soon followed by "First Photographs of the Canals of Mars," *Proceedings of the Royal Society* 77 (Feb. 8, 1906): 132–35.

73. Newcomb to Gill, Jan. 29, 1906, container 52, LOC-SN. Newcomb, in retirement, was still working to compute new lunar tables, funded by the Carnegie Institution.

74. Newcomb to Chisholm, Feb. 19, Mar. 20, 1906; Chisholm to Newcomb, Mar. 7, 30, 1906, containers 17 and 52, LOC-SN. Chisholm, a lawyer and journalist by training, presumably had no special expertise when it came to the question of "Mars," and it is tricky, therefore, to intuit any special agenda or view on his part, beyond an honest desire to assist Newcomb. However, as discussed later in this chapter, some of his interventions would have a major impact on the production of the article.

75. Newcomb's discussions with these scientists also pertained to his related interest in childhood education, which he was attempting to study from the perspective of psychology in the hopes of developing a new system for inculcating good practices of thinking in the nation's youth. As such, Newcomb's interest in the psychology of education shades somewhat into his work on astronomical observation, both being conducted at the same time and both being concerned with the problem of mental perception and inference. See Newcomb's letter to Galton, quoted later in this chapter (ch. 4, n94), as well as that of Nov. 2 [1905] held in UCL-FG. See also all correspondence with Cattell, Carr, and Daniel Coit Gilman in LOC-SN. Around this time Newcomb also requested a collection of publications relating to Mars from the British Astronomical Association. See Newcomb to Chisholm, Nov. 14, 1905, container 52, LOC-SN.

76. Morse, *Mars and its Mystery*, 78; Strauss, "'Fireflies Flashing in Unison'"; Strauss, *Percival Lowell*, 203–19; Markley, *Dying Planet*, 102–5; Lane, *Geographies of Mars*, 100–104, 161–66; Wayman, *Edward Sylvester Morse*.

77. Morse to Newcomb, Nov. 5, 22, 1905; Newcomb to Morse, Nov. 8, 1905, containers 32 and 52, LOC-SN. The debate continued along these lines in Morse to Newcomb, Nov. 24; Newcomb to Morse, Nov. 27, 1905, containers 32 and 52, LOC-SN.

78. Morse to Newcomb, Nov. 24, Dec. 3, 1905, container 32, LOC-SN. Newcomb did concede that Morse's skill as a microscopist would make him "well qualified to catch details made visible with a telescope." See Newcomb's anonymous review of Morse's and Lowell's work in *Nation* 84 (Apr. 4, 1907): 317–18, quote on 317.

79. Newcomb to Morse, Dec. 6, 1905, container 52, LOC-SN.

80. Gill to Newcomb, Mar. 3, 1906; Newcomb to Gill, Apr. 3, 1906, containers 23 and 52, LOC-SN.

81. As such my own assessment of Newcomb's work disagrees with Moyer, *Scientist's Voice in American Culture*, which draws parallels between Newcomb and his pragmatist contemporaries (205–37).

82. Menand, "An Introduction to Pragmatism"; Menand, *Metaphysical Club*; Brent, *Charles Sanders Peirce*; Richardson, *William James*; Simon, *Genuine Reality*; Cadwallader, "Peirce as an Experimental Psychologist"; Lenzen, "Charles S. Peirce as Astronomer"; Hacking, "Telepathy," 431–34; Eisele, *Scientific and Mathematical Philosophy of Charles S. Peirce*; Eisele, *Historical Perspectives on Peirce's Logic*; Bordogna, *William James at the Boundaries*; James, *Principles of Psychology*. On Wundt, see Danziger, "Wilhelm Wundt"; Schaffer, "Astronomers Mark Time."

83. James, *Pragmatism*; James, *Essays in Radical Empiricism*; James to Carl Stumpf, Jan. 1, 1886, as quoted in Simon, *Genuine Reality*, 190 (italics in the original). See also Menand, *Metaphysical Club*, 353; Moyer, *Scientist's Voice in American Culture*, 128–45; Lears, *Rebirth of a Nation*, 238–39; Richardson, *William James*, 193–298.

84. James, "What Psychical Research Has Accomplished," in James, *Will to Believe*, 299. There is a vast literature on James and Spiritualism, including Sommer, "Psychical Research"; Blum, *Ghost Hunters*; Barnard, *Exploring Unseen Worlds*; McDermott, "Introduction." On Peirce's interest in the occult, see Brent, *Charles Sanders Peirce*, 205–318. On the intimate interconnection of new physics, spiritualism, and experimental psychology, see Noakes, "'Cranks and Visionaries'"; Noakes, "Connecting Physical and Psychical Realities."

85. James to Thomas Davidson, Feb. 1, 1885, as quoted in Skrupskelis and Berkeley, *Correspondence of William James*, 6:4.

86. The debate began in public. See James, letter to the editor, "Professor Newcomb's Address before the American Society for Psychical Research," *Science* 7 (Feb. 5, 1886): 123; Newcomb's letter of reply is in the Feb. 12, 1886 issue, 145–46. The two continued in private. See James to Newcomb, Feb. 12, Jul. 7; Newcomb to James, Feb. 16, 1886, as quoted in Skrupskelis and Berkeley, *Correspondence of William James*, 6:115–16, 147. See also Moyer, *Scientist's Voice in American Culture*, 166–82; Barnard, *Exploring Unseen Worlds*, 43–51; Blum, *Ghost Hunters*, 87–90. Many years later, in 1904, James chose to boycott Newcomb's Saint Louis International Congress of Arts and Sciences. See Bordogna, *William James at the Boundaries*, 246–47.

87. Newcomb to Peirce, Mar. 9, 1892, as quoted in Eisele, "Charles S. Peirce–Simon Newcomb Correspondence," 424–25, which reproduces the entire debate. Peirce had contended that "infinite quantities are capable of being conceived as objects of experience, though not of measurement" (416).

88. Peirce to James Baldwin, undated, as quoted in Brent, *Charles Sanders Peirce*, 275.

89. Brent, *Charles Sanders Peirce*, 150–54, 197–202, 288–89.

90. James's published correspondence suggests that he knew the "astronomer and wit" Lowell very well. See Skrupskelis and Berkeley, *Correspondence of William James*, 9:594.

91. Parallels might be drawn here with the German "reactionary moderns" of Herf, *Reactionary Modernism*. Both Coats, "American Scholarship Comes of Age," and Bordogna, *William James at the Boundaries*, 219–58, recognize a Germanic "idealist" hierarchical structure to Newcomb and Münsterberg's 1904 Saint Louis Congress program.

92. Schaffer, "Astronomers Mark Time"; Canales, "Exit the Frog." For the speed of light determination and Newcomb's collaboration with a young Albert Michelson, see Staley, "Conspiracies of Proof"; Staley, *Einstein's Generation*, 27–64; Staley, "Michelson and the Observatory"; Reingold, *Science in Nineteenth Century America*, 275–306; Dick, Orchiston, and Love, "American Transit of Venus Expeditions." On the Venus transit problems, see Ratcliff, *Transit of Venus Enterprise*, 119–71. On French parallels, including Le Verrier's support of Fizeau's research into the determination of the speed of light, see Canales, "Photogenic Venus."

93. Newcomb to Lowell, Feb. 20, 1905, container 52, LOC-SN.

94. Newcomb to Galton, Sep. 29, 1905, container 52, LOC-SN.

95. Newcomb to Lowell, Jan. 12, 1907; Lowell to Newcomb, Jan. 10, 15, 1907, containers 30 and 53, LOC-SN.

96. Newcomb to Chisholm; Newcomb to Edward Pickering, both Jan. 19, 1907, container 53, LOC-SN. Chisholm replied that he didn't like the idea of "mere drawings, which must inevitably 'fake' to some extent the 'canals,'" suggesting that only "reproduction of Lowell's *photographs* would be scientific," but noting that this would be impossible for the *Britannica* to pull off. See Chisholm to Newcomb, Feb. 5, 1907, container 17, LOC-SN.

97. Edward Pickering to Newcomb, Jan. 21, 29, Feb. 16, 26, Mar. 1, 5, 7; Newcomb to Edward Pickering, Jan. 24; Morse to Newcomb, Jan. 31, Feb. 4, Mar. 28; William Pickering to Newcomb, Mar. 16, 22; Frost to Newcomb, May, 27, 28, Jun. 3, 1907, containers 22, 32, 35 and 53, LOC-SN. See also Strauss, *Percival Lowell*, 224–26.

98. Newcomb, "The Optical and Psychological Principles Involved in the Interpretation of the So-Called Canals of Mars," *Astrophysical Journal* 26 (Jul. 1907): 1–17; Barnard to Newcomb, May 31, 1907, container 15, LOC-SN. Barnard also reported that he could replicate the artifactual image of linear canals with photographs of the test image, reiterating, as Frost noted (to Newcomb, Jun. 3, 1907, container 22, LOC-SN), "that photographing a thing is not a final demonstration of its objective reality."

99. Newcomb to Lowell, May 4, 23, 1907, LOA-ECLO; Lowell to Newcomb, May 15, 1907, container 30, LOC-SN.

100. Lowell, "The Canals of Mars, Optically and Psychologically Considered:

A Reply to Professor Newcomb," *Astrophysical Journal* 26 (Oct. 1907): 131–40, 134–35. Newcomb's reply follows on 141, and Lowell's counter-reply on 142.

101. Slipher to Lowell, Sep. 19, 1907, LOA-ECLO (the phrasing is Slipher's); Peirce to Newcomb, Jan. 7, 1908, container 34, LOC-SN.

102. On the open, ambiguous, and finely balanced nature of the canal debate, ca. 1907, see Markley, *Dying Planet*, 100. Newcomb did get in one final counter-punch nine months later, with his "Fallacies about Mars," quoted above in the "Newcomb in the New Age" section of this chapter (ch. 4, n13).

103. Newcomb to Chisholm, Jun. 5, Jul. 3; Chisholm to Newcomb, Jun. 17, 1907, containers 17 and 53, LOC-SN.

104. Throughout the summer of 1907 telegrams sent to the *Herald* by the Lowell Observatory's expedition leader in Chile, David Todd, stirred up a press sensation over reports of greatly improved photographs of the canals. See, for example, "Got Pictures of Canals in Mars," *New York Herald*, Oct. 11, 1907, 9. Lowell eventually instigated a bidding war for the rights to print exclusive reproductions upon Todd's return. See Lane, *Geographies of Mars*, 115–28; Tucker, *Nature Exposed*, 224–30.

105. Newcomb, "Mars," *Encyclopaedia Britannica*, 11th ed., vol. 17, *Lord Chamberlain to Mecklenburg* (Cambridge: Cambridge University Press, 1911), 761–65. Lowell's footnote can be found on 764. The maps of Mars provided by the Lowell Observatory are on 764 and 765. A footnote was also added to Newcomb's explanation of the potential for erroneous visual inference in canal observations, directing readers to Lowell's rebuttal.

106. Newcomb's own actions and intentions are, therefore, easily misunderstood by readers unfamiliar with the *Britannica* article's particular context of production. For example, the historian of science fiction Paul Fayter has recently written that "in his 1911 article on Mars for the eleventh edition of the *Encyclopaedia Britannica*, Newcomb helped enshrine a Lowellian interpretation of Mars by including two drawings of the Martian canal system." See n5 in Fayter, "'Some Eden Lost in Space.'"

107. For example, Oreskes, "Rejection of Continental Drift"; Schaffer, "Where Experiments End."

CONCLUSION

1. Lowell, "Mars and the Earth," lecture text, ca. Aug. 1916, as quoted in Hoyt, *Lowell and Mars*, 300. On the poor quality of Martian oppositions after 1909, see the table in Sheehan, *Planet Mars*, 227.

2. Detailed coverage of the 1909 opposition and its fallout is given in Crowe, *Extraterrestrial Life Debate*, 524–46; Sheehan, *Planet Mars*, 130–45; Hoyt, *Lowell and Mars*, 287–302; Dick, *Biological Universe*, 79–105; Strauss, *Percival Lowell*, 220–40; Lane, *Geographies of Mars*, 58–62.

3. Campbell to Hale, May 11, 1908, as quoted in Sheehan, *Planet Mars*, 131–32; Hale to Campbell, May 20, 1908, as quoted in DeVorkin, "Campbell's Spectroscopic Study," 43.

4. "Deny Mars has Canals," *New York Times*, Dec. 30, 1909, 1.

5. Antoniadi, "Fourth Interim Report for the Apparition of 1909 . . . the Planet Mars," *Journal of the British Astronomical Association* 20 (Nov. 1909): 78–81, 79; "Report of the Meeting of the Association . . . Dec. 29, 1909," *Journal of the British Astronomical Association* 20 (Dec. 1909): 119–28, 123 (emphasis mine).

6. Planetary science as a subdiscipline likely suffered as a result of this new-found unity. As Iosef Shklovskii and Carl Sagan (somewhat self-servingly) noted in their *Intelligent Life in the Universe*, the "bitter" fallout over Martian canals "led to a general exodus from planetary to stellar astronomy" (276).

7. Markley, *Dying Planet*, 113.

8. Frost, *Astronomer's Life*, 217. Carl Sagan would embellish this story some-what in his blockbuster *Cosmos*. In Sagan's recounting, the request comes from a "celebrated newspaper publisher" asking "a noted astronomer" for five hundred words on whether there is life on Mars. "The astronomer dutifully replied: NOBODY KNOWS, NOBODY KNOWS, NOBODY KNOWS . . . 250 times" (106). See also Crowe, "Inflation and History."

9. Hofling, "Lowell and the Canals of Mars," 40.

10. Sagan and Fox, "Canals of Mars," 602. A decade earlier, hopes of just such an explanation had been higher. In the immediate wake of Mariner 4's first fly-by photographs of the Martian surface, Sagan and James Pollock had argued for the canals as corresponding to real geological features. See Markley, *Dying Planet*, 180. Disciplinary embarrassment over Mars and subsequent boundary work by planetary scientists is discussed in: Doel, *Solar System Astronomy in America*, 13–15; Smith, "Martians and Other Aliens," 244–45; Tatarewicz, *Space Technology and Planetary Astronomy*, 2–6.

11. Shklovskii and Sagan, *Intelligent Life in the Universe*, 276.

12. Key examples include Asimov, *Guide to Science*, 185–86; Heffernan, "The Singularity of Our Inhabited World"; Hetherington, "Percival Lowell"; Gould, "War of the Worldviews"; Pillinger, *Beagle*, 34–57; Sobel, *Planets*, 134–35.

13. Markley, *Dying Planet*, 150–81. Further detail on this era can be found in Dick, *Biological Universe*, 99–126; Sheehan, *Planet Mars*, 146–61; Doel, *Solar System Astronomy*, 9–22; Tatarewicz, *Space Technology and Planetary Astronomy*, 1–6.

14. Bensaude-Vincent, "Science and the Public"; Bowler, *Science for All*.

15. Crelinsten, *Einstein's Jury*.

16. Molvig, "Cosmological Revolutions"; Staley, *Einstein's Generation*; Staley, "Co-Creation of Classical and Modern Physics"; Sponsel, "Constructing a 'Revolution in Science'"; Clarke, "How to Manage a Revolution."

17. Silas Bent, "Mars Invites Mankind to Reveal His Secret," *New York Times*, Aug. 17, 1924, sec. 8, 6; "Did Mars Speak to Us? Canadian Radio Men Agog," *Chicago Daily Tribune*, Aug. 22, 1924, 3; "Eyes of World on Mars as Earth Passes Planet," *Baltimore Sun*, Aug. 23, 1924, 7; "Weird 'Radio Signal' Film Deepens Mystery of Mars," *Washington Post*, Aug. 27, 1924, 2; "Camera Film in Mars Mystery," *Los Angeles Times*, Aug. 28, 1924, 4; "78 Army and Navy Radios Fail to Hear Mars Signal," *Washington Post*, Aug. 29, 1924, 13; F. B. C., letter to the editor, "Signaling to Mars," *Christian Science Monitor*, Sep. 16, 1924, 20. Interest in radio signals from Mars at this time was stoked in part by Guglielmo Marconi's keen interest. See Dick, *Biological Universe*, 401–10.

BIBLIOGRAPHY

ARCHIVAL SOURCES

Adams, John Couch, Papers. GBR/0275/Adams. Saint John's College Library, Cambridge, UK.

Draper, Henry and Mary Anna Palmer, Papers, 1859–1914. MssCol 838. New York Public Library, New York.

Early Correspondence of the Lowell Observatory, 1894–1916. Microfilm Edition. Lowell Observatory Archive, Flagstaff, AZ.

Galton, Sir Francis, Papers and Correspondence. GB 0103 GALTON. University College Library, Special Collections, London, UK.

Lockyer, Sir Norman, Research Papers. EUL MS 110. Special Collections, University of Exeter Library, Exeter, UK.

Newcomb, Simon, Papers, 1854–1936. MSS34629. Library of Congress, Washington, DC.

Pickering, Edward C., Director, Harvard College Observatory, Records, 1864–1926, HOLLIS No. 001966647. Harvard University Archives, Cambridge, MA.

Pickering, William H., Papers, 1868–1916. HOLLIS No. 000604354. Harvard University Archives, Cambridge, MA.

Proctor, Richard Anthony, Letters. PR2—Miscellaneous Papers. Saint John's College Library, Cambridge, UK.

REFERENCES

Encyclopedias to which Simon Newcomb contributed are catalogued separately after this list.

Adams, Henry. *The Education of Henry Adams*. Boston: Houghton Mifflin, 1918.

Adas, Michael. *Dominance by Design: Technological Imperatives and America's Civilizing Mission*. Cambridge, MA: Belknap Press, 2006.

Ahvenainen, Jorma. *The European Cable Companies in South America*. Helsinki: Academia Scientiarum Fennica, 2004.

Alberti, Samuel. "Amateurs and Professionals in One County: Biology and Natural History in Late Victorian Yorkshire." *Journal of the History of Biology* 34, no. 1 (2001): 115–47.

Altick, Richard D. *The English Common Reader: A Social History of the Mass Reading Public, 1800–1900*. Chicago: University of Chicago Press, 1963.

Altick, Richard D. *The Shows of London*. Cambridge, MA: Harvard University Press, 1978.

Anderson, Katherine. *Predicting the Weather: Victorians and the Science of Meteorology*. Chicago: University of Chicago Press, 2005.

Archibald, Raymond C. "Simon Newcomb, 1835–1909: Bibliography of his Life and Work." *Memoirs of the National Academy of Sciences* 17 (1924): 19–69.

Ashworth, William J. "The Calculating Eye: Baily, Herschel, Babbage and the Business of Astronomy." *British Journal for the History of Science* 27, no. 4 (1994): 409–41.

Ashworth, William J. "John Herschel, George Airy and the Roaming Eye of the State." *History of Science* 36, no. 2 (1998): 151–78.

Asimov, Isaac. *Asimov's Guide to Science*. Vol. 2, *The Biological Sciences*. Harmondsworth: Penguin, 1975.

Aubin, David, and Charlotte Bigg. "Neither Genius nor Context Incarnate: Lockyer, Janssen and the Astrophysical Self." In *The History and Poetics of Scientific Biography*, edited by Thomas Söderqvist, 51–70. Aldershot: Ashgate, 2007.

Aubin, David, Charlotte Bigg, and H. Otto Sibum, eds. *The Heavens on Earth: Observatories and Astronomy in Nineteenth-Century Science and Culture*. Durham, NC: Duke University Press, 2010.

Badash, Lawrence. "The Completeness of Nineteenth-Century Science." *Isis* 63, no. 1 (1972): 48–58.

Baldwin, Melinda. "The Shifting Ground of Nature: Establishing an Organ of Scientific Communication in Britain, 1869–1900." *History of Science* 50, no. 2 (2012): 125–54.

Baldwin, Melinda. "The Successors to the X Club? Late Victorian Naturalists and *Nature*, 1869–1900." In *Victorian Scientific Naturalism: Community, Identity, Continuity*, edited by Gowan Dawson and Bernard Lightman, 288–308. Chicago: University of Chicago Press, 2014.

Baldwin, Melinda. *Making "Nature": The History of a Scientific Journal*. Chicago: University of Chicago Press, 2015.

Ball, Robert S. *In Starry Realms*. London: Isbister, 1892.

Barber, William J. "Should the American Economic Association Have Toasted Simon Newcomb at Its 100th Birthday Party?" *Journal of Economic Perspectives* 1, no. 1 (1987): 179–83.

Barnard, George William. *Exploring Unseen Worlds: William James and the Philosophy of Mysticism*. New York: State University of New York Press, 1997.

Bartky, Ian R. *Selling the True Time: Nineteenth-Century Timekeeping in America*. Stanford, CA: Stanford University Press, 2000.

Barton, Ruth. "'An Influential Set of Chaps': The X-Club and Royal Society Politics, 1864–85." *British Journal for the History of Science* 23, no. 1 (1990): 53–81.

Barton, Ruth. "'Huxley, Lubbock, and Half a Dozen Others': Professionals and Gentlemen in the Formation of the X Club, 1851–1864." *Isis* 89, no. 3 (1998): 410–44.

Barton, Ruth. "Just Before Nature: The Purposes of Science and the Purposes of Popularization in Some English Popular Science Journals of the 1860s." *Annals of Science* 55, no. 1 (1998): 1–33.

Barton, Ruth. "'Men of Science': Language, Identity and Professionalization in the Mid-Victorian Scientific Community." *History of Science* 41, no. 1 (2003): 73–119.

Barton, Ruth. "Sunday Lecture Societies: Naturalistic Scientists, Unitarians, and Secularists Unite against Sabbatarian Legislation." In *Victorian Scientific Naturalism: Community, Identity, Continuity*, edited by Gowan Dawson and Bernard Lightman, 189–219. Chicago: University of Chicago Press, 2014.

Batten, Alan H. *Resolute and Undertaking Characters: The Lives of Wilhelm and Otto Struve*. Dordecht: Reidel, 1988.

Baylen, Joseph O. "W. T. Stead and the 'New Journalism.'" *Emory University Quarterly* 21, no. 3 (1965): 1–13.

Baylen, Joseph O. "The 'New Journalism' in Late Victorian Britain." *Australian Journal of Politics and History* 18, no. 3 (1972): 367–85.

Baylen, Joseph O. "W. T. Stead as Publisher and Editor of the 'Review of Reviews.'" *Victorian Periodicals Review* 12, no. 2 (1979): 70–84.

Baylen, Joseph O. "Stead, William Thomas (1849–1912)." In *Biographical Dictionary of Modern British Radicals*, edited by Joseph O. Baylen and Norbert J. Gossman, 783–92. Brighton: Harvester, 1988.

Beach, Mark. "Was There a Scientific Lazzaroni?" In *Nineteenth-Century American Science: A Reappraisal*, edited by George H. Daniels, 115–32. Evanston, IL: Northwestern University Press, 1972.

Beck, Peter J. *The War of the Worlds: From H. G. Wells to Orson Welles, Jeff Wayne, Steven Spielberg and Beyond*. London: Bloomsbury Academic, 2016.

Becker, Barbara J. *Unravelling Starlight: William and Margaret Huggins and the Rise of the New Astronomy*. Cambridge: Cambridge University Press, 2011.

Belknap, Geoffrey D. *From a Photograph: Authenticity, Science and the Periodical Press, 1870–1890*. London: Bloomsbury, 2016.

Bell, Duncan. "The Project for a New Anglo Century: Race, Space, and Global Order." In *Anglo-America and Its Discontents: Civilizational Politics beyond East and West*, edited by Peter J. Katzenstein, 33–55. Abingdon: Routledge, 2012.

Bell, Duncan. "Dreaming the Future: Anglo-America as Utopia, 1880–1914." In *The American Experiment and the Idea of Democracy in British Culture, 1776–1914*, edited by Ella Dzelzainis and Ruth Livesey, 197–210. Aldershot: Ashgate, 2013.

Bell, Duncan. *Dreamworlds of Race: Empire, Utopia, and the Fate of Anglo-America*. Princeton: Princeton University Press, forthcoming.

Bellon, Richard. "Science at the Crystal Focus of the World." In *Science in the Marketplace: Nineteenth-Century Sites and Experiences*, edited by Aileen Fyfe and Bernard Lightman, 301–35. Chicago: University of Chicago Press, 2007.

Bennett, Jim. "The Era of Newton, Herschel and Lord Rosse." *Experimental Astronomy* 25, no. 1–3 (2009): 33–42.

Bennett, Jim, Robert Brain, Simon Schaffer, Heinz Otto Sibum, and Richard Staley. *1900: The New Age, A Guide to the Exhibition*. Cambridge: Whipple Museum of the History of Science, 1994.

Bensaude-Vincent, Bernadette. "Camille Flammarion: Prestige de la Science Populaire." *Romantisme* 19, no. 65 (1989): 93–104.

Bensaude-Vincent, Bernadette. "A Genealogy of the Increasing Gap Between Science and the Public." *Public Understanding of Science* 10, no. 1 (2001): 99–113.

Bensaude-Vincent, Bernadette. "A Historical Perspective on Science and Its 'Others.'" *Isis* 100, no. 2 (2009): 359–68.

Berthold, Victor M. *History of the Telephone and Telegraph in Chile, 1851–1921*. New York: American Telephone and Telegraph Company, 1924.

Bigg, Charlotte. "Spectroscopic Metrologies." *Nuncius* 18, no. 2 (2003): 765–77.

Bigg, Charlotte. "Staging the Heavens: Astrophysics and Popular Astronomy in the Late Nineteenth Century." In *The Heavens on Earth: Observatories and Astronomy in Nineteenth-Century Science and Culture*, edited by David Aubin, Charlotte Bigg, and H. Otto Sibum, 305–24. Durham, NC: Duke University Press, 2010.

Bigg, Charlotte. "Travelling Scientist, Circulating Images and the Making of the Modern Scientific Journal: Norman Lockyer's Visual Communication of Astrophysics in *Nature*." *Nuncius* 30, no. 3 (2015): 675–98.

Bigg, Charlotte, and Kurt Vanhoutte. "Spectacular Astronomy." *Early Popular Visual Culture* 15, no. 2 (2017): 115–24.

Blondheim, Menahem. *News over the Wires: The Telegraph and the Flow of Public Information in America, 1844–1897*. Cambridge, MA: Harvard University Press, 1994.

Blum, Deborah. *Ghost Hunters: William James and the Search for Scientific Proof of Life after Death.* New York: Penguin, 2007.

Bode, Carl. *The American Lyceum: Town Meeting of the Mind.* Oxford: Oxford University Press, 1956.

Bordogna, Francesca. *William James at the Boundaries: Philosophy, Science and the Geography of Knowledge.* Chicago: University of Chicago Press, 2008.

Boston, Ray. "W. T. Stead and Democracy by Journalism." In *Papers for the Millions: The New Journalism in Britain, 1850s to 1914,* edited by Joel H. Wiener, 91–106. New York: Greenwood Press, 1988.

Bould, Mark, and China Miéville, eds. *Red Planets: Marxism and Science Fiction.* London: Pluto Press, 2009.

Bowler, Peter. *Science for All: The Popularization of Science in Early Twentieth-Century Britain.* Chicago: University of Chicago Press, 2009.

Boyd, James P. *The Paris Exposition of 1900: A Vivid Descriptive View and Elaborate Scenic Presentation of the Site, Plan and Exhibits.* Philadelphia: P. W. Ziegler, 1900.

Boyd, Lyle G. "Mrs. Henry Draper and the Harvard College Observatory: 1883–1887." *Harvard Library Bulletin* 17, no. 1 (1969): 70–97.

Boyle, Colleen. "You Saw the Whole of the Moon: The Role of Imagination in the Perceptual Construction of the Moon." *Leonardo* 46, no. 3 (2013): 246–52.

Brain, Robert. *Going to the Fair: Readings in the Culture of Nineteenth-Century Exhibitions.* Cambridge: Whipple Museum of the History of Science, 1993.

Brain, Robert. "Going to the Exhibition." In *The Physics of Empire: Public Lectures,* edited by Richard Staley, 113–42. Cambridge: Whipple Museum of the History of Science, 1994.

Brake, Laurel. "The Old Journalism and the New: Forms of Cultural Production in London in the 1880s." In *Papers for the Millions: The New Journalism in Britain, 1850s to 1914,* edited by Joel H. Wiener, 1–24. New York: Greenwood Press, 1988.

Brake, Laurel. "Writing Women's History: 'The Sex' Debates of 1889." In *New Woman Hybridities: Feminism, Femininity and International Consumer Culture, 1880–1930,* edited by Ann Heilmann and Margaret Beetham, 51–73. London: Routledge, 2004.

Brake, Laurel, Aled Jones, and Lionel Madden, eds. *Investigating Victorian Journalism.* Basingstoke: Macmillan, 1990.

Brent, Joseph. *Charles Sanders Peirce: A Life.* Bloomington: Indiana University Press, 1998.

Brock, William H. "Science, Technology and Education in *The English Mechanic.*" In *Science for All: Studies in the History of Victorian Science and Education,* by William H. Brock. Aldershot: Variorum, 1996.

Brooke, John H. "Natural Theology and the Plurality of Worlds: Observations on the Brewster-Whewell debate." *Annals of Science* 34, no. 3 (1977): 221–86.

Broks, Peter. *Media Science before the Great War.* Basingstoke: Macmillan, 1996.

Brown, Lucy. "The Treatment of the News in Mid-Victorian Newspapers." *Transactions of the Royal Historical Society*, 5th ser., 27 (1977): 23–39.

Brown, Lucy. *Victorian News and Newspapers.* Oxford: Clarendon, 1985.

Bruce, Robert V. *The Launching of Modern American Science, 1846–1876.* New York: Knopf, 1987.

Brück, Mary. *Agnes Mary Clerke and the Rise of Astrophysics.* Cambridge: Cambridge University Press, 2008.

Brush, Stephen G. "Looking Up: The Rise of Astronomy in America." *American Studies* 20, no. 2 (1979): 41–67.

Brush, Stephen G. "The Nebular Hypothesis and the Evolutionary Worldview." *History of Science* 25, no. 3 (1987): 245–78.

Brush, Stephen G. *A History of Modern Planetary Physics.* 3 vols. Cambridge: Cambridge University Press, 1996.

Burchfield, Joe D. *Lord Kelvin and the Age of the Earth.* New York: Science History Publications, 1975.

Burkhardt, Frederick, and James Secord, eds. *The Correspondence of Charles Darwin.* Vol. 19, *1871.* Cambridge: Cambridge University Press, 2012.

Burnett, John. "British Studies of Mars: 1877–1914." *Journal of the British Astronomical Association* 89, no. 2 (1979): 136–43.

Bush, Martin. "The Proctor-Parkes Incident: Politics, Protestants and Popular Astronomy in Australia in 1880." *Historical Records of Australian Science* 28, no. 1 (2017): 26–36.

Butterworth, Mark. "Astronomical Lantern Slides." *New Magic Lantern Journal* 10, no. 4 (2008), 65–68.

Byrn, Edward W. *The Progress of Invention in the Nineteenth Century.* New York: Munn, 1900.

Cadwallader, Thomas C. "Peirce as an Experimental Psychologist." *Transactions of the Charles S. Peirce Society* 11, no. 3 (1975): 167–86.

Campbell, Kate. "W. E. Gladstone, W. T. Stead, Matthew Arnold and a New Journalism: Cultural Politics in the 1880s." *Victorian Periodicals Review* 36, no. 1 (2003): 20–40.

Canadelli, Elena. "'Some Curious Drawings': Mars through Giovanni Schiaparelli's Eyes, Between Science and Fiction." *Nuncius* 24, no. 2 (2009): 439–64.

Canales, Jimena. "Exit the Frog, Enter the Human: Physiology and Experimental Psychology in Nineteenth-Century Astronomy." *British Journal for the History of Science* 34, no. 2 (2001): 173–97.

Canales, Jimena. "Photogenic Venus: The 'Cinematographic Turn' and Its Alternatives in Nineteenth-Century France." *Isis* 93, no. 4 (2002): 585–613.

Cannon, Walter F. "John Herschel and the Idea of Science." *Journal of the History of Ideas* 22, no. 2 (1961): 215–39.

Cantor, Geoffrey, Gowan Dawson, Graeme Gooday, Richard Noakes, Sally Shuttleworth, and Jonathan R. Topham. *Science in the Nineteenth-Century Periodical: Reading the Magazine of Nature.* Cambridge: Cambridge University Press, 2004.

Cantor, Geoffrey, and Sally Shuttleworth, eds. *Science Serialized: Representation of the Sciences in Nineteenth-Century Periodicals.* Cambridge, MA: MIT Press, 2004.

Carey, James W. *The Intellectuals and the Masses: Pride and Prejudice among the Literary Intelligentsia, 1880–1930.* London: Faber and Faber, 1992.

Carey, James W. *Communication as Culture: Essays on Media and Society.* New York: Routledge, 2009.

Carlson, W. Bernard. *Tesla: Inventor of the Electrical Age.* Princeton: Princeton University Press, 2013.

Carson, Cathryn. *Heisenberg in the Atomic Age: Science and the Public Sphere.* New York: Cambridge University Press, 2010.

Carter, Bill, and Merri Sue Carter. *Simon Newcomb: America's Unofficial Astronomer Royal.* Saint Augustine, FL: Mantanzas Publishing, 2006.

Carver, Ben. "'A Gleaming and Glorious Star': Rethinking History in the Plurality-of-Worlds Debate." *Journal of Victorian Culture* 18, no. 4 (2013): 429–52.

Chalaby, Jean K. "Journalism as an Anglo-American Invention: A Comparison of the Development of French and Anglo-American Journalism, 1830s–1920s." *European Journal of Communication* 11, no. 3 (1996): 303–26.

Chalaby, Jean K. *The Invention of Journalism.* Basingstoke: Macmillan, 1998.

Channing, William E. *The Works of William E. Channing, D.D.* Boston: American Unitarian Association, 1890.

Chapman, Allan. "Private Research and Public Duty: George Biddell Airy and the Search for Neptune." *Journal for the History of Astronomy* 19, no. 2 (1988): 121–39.

Chapman, Allan. "William Lassell (1799–1880): Practitioner, Patron and 'Grand Amateur' of Victorian Astronomy." *Vistas in Astronomy* 32, no. 4 (1988): 341–70.

Chapman, Allan. "George Biddell Airy, FRS (1801–1892): A Centenary Commemoration." *Notes and Records of the Royal Society* 46, no. 1 (1992): 103–10.

Chapman, Allan. *The Victorian Amateur Astronomer: Independent Astronomical Research in Britain, 1820–1920.* New York: Wiley, 1998.

Churchill, Allen. *Park Row.* New York: Rinehart, 1958.

Clarke, Imogen. "How to Manage a Revolution: Isaac Newton in the Early Twentieth Century." *Notes and Records of the Royal Society* 68 (2014): 323–37.

Clements, William M. "Overland Monthly." In *American Literary Magazines: The Eighteenth and Nineteenth Centuries*, edited by Edward E. Chielens, 308–13. New York: Greenwood Press, 1986.

Clerke, Agnes M. *A Popular History of Astronomy during the Nineteenth Century.* Edinburgh: A. and C. Black, 1885.

Clerke, Agnes M. *The System of the Stars.* London: Longmans, Green, 1890.

Clifford, David, Elisabeth Wadge, Alex Warwick, and Martin Willis, eds. *Repositioning Victorian Sciences: Shifting Centres in Nineteenth-Century Scientific Thinking.* London: Anthem Press, 2006.

Coats, A. W. "American Scholarship Comes of Age: The Louisiana Purchase Exposition 1904." *Journal of the History of Ideas* 22, no. 3 (1961): 404–17.

Colligan, Colette, and Margaret Linley. "Introduction: The Nineteenth-Century Invention of Media." In *Media, Technology, and Literature in the Nineteenth Century: Image, Sound, Touch*, edited by Colette Colligan and Margaret Linley. Aldershot: Ashgate, 2011.

Collini, Stefan. *Public Moralists: Political Thought and Intellectual Life in Britain, 1850–1930.* Oxford: Clarendon, 1991.

Collins, Harry M. *Changing Order: Replication and Induction in Scientific Practice.* Chicago: University of Chicago Press, 1992.

Collins, Paul. *The Murder of the Century: The Gilded Age Crime That Scandalized a City and Sparked the Tabloid Wars.* New York: Broadway Paperbacks, 2011.

Collison, Robert L. *Encyclopaedias: Their History throughout the Ages.* New York: Hafner, 1966.

Conway, Jill. "Stereotypes of Femininity in a Theory of Sexual Evolution." In *Suffer and Be Still: Women in the Victorian Age*, edited by Martha Vicinus, 140–154. Bloomington: Indiana University Press, 1972.

Cooter, Roger. *The Cultural Meaning of Popular Science: Phrenology and the Organization of Consent in Nineteenth-Century Britain.* Cambridge: Cambridge University Press, 1984.

Cooter, Roger, and Stephen Pumfrey. "Separate Spheres and Public Places: Reflections on the History of Science Popularization and Science in Popular Culture." *History of Science* 32, no. 3 (1994): 237–67.

Cottam, Stella, and Wayne Orchiston. *Eclipses, Transits, and Comets of the Nineteenth Century: How America's Perception of the Skies Changed.* New York: Springer, 2015.

Crelinsten, Jeffrey. *Einstein's Jury: The Race to Test Relativity.* Princeton: Princeton University Press, 2006.

Cronon, William. *Nature's Metropolis: Chicago and the Great West.* New York: W. W. Norton, 1991.

Crossley, Robert. "Percival Lowell and the History of Mars." *Massachusetts Review* 41, no. 3 (2000): 297–318.

Crossley, Robert. "H. G. Wells, Visionary Telescopes, and the 'Matter of Mars.'" *Philological Quarterly* 83, no. 1 (2004): 83–114.

Crossley, Robert. "The Grandeur of H. G. Wells." In *A Companion to Science Fiction*, edited by David Seed, 353–63. Oxford: Blackwell, 2005.

Crossley, Robert. "Mars and the Paranormal." *Science Fiction Studies* 35, no. 3 (2008): 466–84.

Crossley, Robert. *Imagining Mars: A Literary History*. Middletown, CT: Wesleyan University Press, 2011.

Crouthamel, James L. *Bennett's "New York Herald" and the Rise of the Popular Press*. Syracuse, NY: Syracuse University Press, 1989.

Crowe, Michael J. "Inflation and History: E. B. Frost's Mars Telegram." *Griffith Observer* 46, no. 3 (1982): 15.

Crowe, Michael J. *The Extraterrestrial Life Debate, 1750–1900: The Idea of a Plurality of Worlds from Kant to Lowell*. Cambridge: Cambridge University Press, 1986.

Crowe, Michael J. "Proctor, Richard Anthony (1837–1888)." In *History of Astronomy: An Encyclopedia*, edited by John Lankford, 416–17. New York: Garland, 1997.

Crowe, Michael J. "Astronomy and Religion (1780–1915): Four Case Studies Involving Ideas of Extraterrestrial Life." *Osiris* 16 (2001): 209–26.

Crowe, Michael J. "William and John Herschel's Quest for Extraterrestrial Intelligent Life." In *The Scientific Legacy of William Herschel*, edited by Clifford J. Cunningham, 239–74. Cham: Springer, 2018.

Csiszar, Alex. "Seriality and the Search for Order: Scientific Print and its Problems During the Late Nineteenth Century." *History of Science* 48, no. 3–4 (2010): 399–434.

Csiszar, Alex. "How Lives Became Lists and Scientific Papers Became Data: Cataloguing Authorship during the Nineteenth Century." *British Journal for the History of Science* 50, no. 1 (2017): 23–60.

Csiszar, Alex. *The Scientific Journal: Authorship and the Politics of Knowledge in the Nineteenth Century*. Chicago: University of Chicago Press, 2018.

Curran, James. "Capitalism and Control of the Press, 1800–1975." In *Mass Communication and Society*, edited by James Curran, Michael Gurevitch and Janet Woollacott, 195–230. London: Edward Arnold, 1977.

Curran, James, and Jean Seaton. *Power without Responsibility: The Press and Broadcasting in Britain*. London: Routledge, 1991.

Curti, Merle. *The Growth of American Thought*. New Brunswick, NJ: Transaction, 1984.

Danziger, K. "Wilhelm Wundt and the Emergence of Experimental Psychology." In *Companion to the History of Modern Science*, edited by Robert C. Olby, Geoffrey N. Cantor, John R. R. Christie, and M. Jonathan S. Hodge, 396–409. London: Routledge, 1990.

Darnton, Robert. *The Business of Enlightenment: A Publishing History of the Encyclopédie, 1775–1800.* Cambridge, MA: Harvard University Press, 1979.

Darnton, Robert. "An Early Information Society: News and the Media in Eighteenth-Century Paris." *American Historical Review* 105, no. 1 (2000): 1–35.

Daston, Lorraine. "Objectivity and the Escape from Perspective." *Social Studies of Science* 22, no. 4 (1992): 597–618.

Daston, Lorraine. "Fear and Loathing of the Imagination in Science." *Daedalus* 127, no. 1 (1998): 73–95.

Daston, Lorraine, and Peter Galison. "The Image of Objectivity." *Representations* 40 (1992): 81–128.

Daston, Lorraine, and Peter Galison. *Objectivity.* New York: Zone Books, 2007.

Davis, Mike. *Late Victorian Holocausts: El Niño Famines and the Making of the Third World.* London: Verso, 2001.

Dawson, Gowan. "The *Review of Reviews* and the New Journalism in Late-Victorian Britain." In *Science in the Nineteenth-Century Periodical: Reading the Magazine of Nature*, by Geoffrey Cantor, Gowan Dawson, Graeme Gooday, Richard Noakes, Sally Shuttleworth, and Jonathan R. Topham, 172–195. Cambridge: Cambridge University Press, 2004.

Dawson, Gowan. "Stead, William Thomas." In *Dictionary of Nineteenth-Century Journalism in Great Britain and Ireland*, edited by Laurel Brake and Marysa Demoor, 598. London: Academia Press, 2009.

Dawson, Gowan, and Bernard Lightman, eds. *Victorian Scientific Naturalism: Community, Identity, Continuity.* Chicago: University of Chicago Press, 2014.

Derrida, Jacques. "Archive Fever: A Freudian Impression." *Diacritics* 25 (1995): 9–63.

Desmond, Adrian. *Thomas Huxley: From Devil's Disciple to Evolution's High Priest.* London: Penguin, 1998.

Desmond, Adrian. "Redefining the X Axis: 'Professionals,' 'Amateurs' and the Making of Mid-Victorian Biology—A Progress Report." *Journal of the History of Biology* 34, no. 1 (2001): 3–50.

DeVorkin, David H. "W. W. Campbell's Spectroscopic Study of the Martian Atmosphere." *Quarterly Journal of the Royal Astronomical Society* 18, no. 1 (1977): 37–53.

DeVorkin, David H., ed. *The American Astronomical Society's First Century.* Washington, DC: American Institute of Physics, 1999.

DeVorkin, David H. *Henry Norris Russell: Dean of American Astronomers.* Princeton: Princeton University Press, 2000.

DeVorkin, David H. "In the Grip of the Big Telescope Age." *Experimental Astronomy* 25, no. 1–3 (2009): 63–77.

Dewhirst, David W. "The Greenwich-Cambridge Axis." *Vistas in Astronomy* 20 (1976): 109–11.

Diamond, B. I. "A Precursor of the New Journalism: Frederick Greenwood of the *Pall Mall Gazette.*" In *Papers for the Millions: The New Journalism in Britain, 1850s to 1914*, edited by Joel H. Wiener, 25–45. New York: Greenwood Press, 1988.

Dick, Steven J. *Plurality of Worlds: The Origins of the Extraterrestrial Life Debate from Democritus to Kant.* Cambridge: Cambridge University Press, 1982.

Dick, Steven J. *The Biological Universe: The Twentieth-Century Extraterrestrial Life Debate and the Limits of Science.* Cambridge: Cambridge University Press, 1996.

Dick, Steven J. *Sky and Ocean Joined: The US Naval Observatory, 1830–2000.* Cambridge: Cambridge University Press, 2003.

Dick, Steven J., Wayne Orchiston, and Tom Love. "Simon Newcomb, William Harkness and the Nineteenth-Century American Transit of Venus Expeditions." *Journal for the History of Astronomy* 29, no. 3 (1998): 221–55.

Doel, Ronald E. *Solar System Astronomy in America: Communities, Patronage, and Interdisciplinary Science, 1920–1960.* Cambridge: Cambridge University Press, 1996.

Donnelly, Kevin. "On the Boredom of Science: Positional Astronomy in the Nineteenth Century." *British Journal for the History of Science* 47, no. 3 (2014): 479–503.

Drew, John. "The Nineteenth-Century Commercial Traveller and Dickens's 'Uncommercial' Philosophy." Part 1, *Dickens Quarterly* 15, no. 1 (1998): 50–61; Part 2, *Dickens Quarterly* 15, no. 2 (1998): 83–110.

Easley, Alexis. "W. T. Stead, Late Victorian Feminism, and the *Review of Reviews.*" In *W. T. Stead: Newspaper Revolutionary*, edited by Roger Luckhurst, Laurel Brake, James Mussell, and Ed King, 37–58. London: British Library, 2012.

Eckley, Grace. *Maiden Tribute: A Life of W. T. Stead.* Philadelphia: Xlibris, 2007.

Einbinder, Harvey. *The Myth of the Britannica.* London: MacGibbon and Kee, 1964.

Eisele, Carolyn. "The Charles S. Peirce–Simon Newcomb Correspondence." *Proceedings of the American Philosophical Society* 101, no. 5 (1957): 409–33.

Eisele, Carolyn. *Studies in the Scientific and Mathematical Philosophy of Charles S. Peirce.* The Hague: Mouton, 1979.

Eisele, Carolyn, ed. *Historical Perspectives on Peirce's Logic of Science: A History of Science.* 2 vols. Berlin: Mouton, 1985.

Ellis, Heather. *Masculinity and Science in Britain, 1831–1918.* London: Palgrave Macmillan, 2017.

Ellis, L. Ethan. "Print Paper Pendulum: Group Pressures and the Price of Newsprint." Appendix in *Newsprint: Producers, Publishers, Political Pressures.* New Brunswick, NJ: Rutgers University Press, 1960.

Endersby, Jim. "Odd Man Out: Was Joseph Hooker an Evolutionary Naturalist?" In *Victorian Scientific Naturalism: Community, Identity, Continuity*, edited by Gowan Dawson and Bernard Lightman, 157–85. Chicago: University of Chicago Press, 2014.

Erickson, Emily. "Yellow Journalism." In *Encyclopedia of American Journalism*, edited by Stephen L. Vaughn, 607–8. New York: Routledge, 2008.

Escott, T. H. S. *Masters of English Journalism: A Study of Personal Forces*. London: T. Fisher Unwin, 1911.

Ewalts, Anouk. "*English Mechanic* (1865–1926)." In *Dictionary of Nineteenth-Century Journalism in Great Britain and Ireland*, edited by Laurel Brake and Marysa Demoor, 204–5. London: Academia Press, 2009.

Fayter, Paul. "Strange New Worlds of Space and Time: Late Victorian Science and Science Fiction." In *Victorian Science in Context*, edited by Bernard Lightman, 256–80. Chicago: University of Chicago Press, 1997.

Fayter, Paul. "'Some Eden Lost in Space': The Wider Contexts of Frederick Philip Grove's 'The Legend of the Planet Mars' (1915)." Conference paper presented at Science Fiction: The Interdisciplinary Genre, McMaster University, Hamilton, Ontario, Canada, September 2013. http://hdl.handle.net/11375/23227.

Fernie, Donald. *The Whisper and the Vision: The Voyages of the Astronomers*. Toronto: Clarke, Irwin, 1976.

Fettmann, Eric. "*New York World.*" In *Encyclopedia of American Journalism*, edited by Stephen L. Vaughn, 345–46. New York: Routledge, 2008.

Flammarion, Camille. *Popular Astronomy: A General Description of the Heavens*. London: Chatto and Windus, 1894. First published in French in numerous editions from 1880.

Flint, Kate. *The Victorians and the Visual Imagination*. Cambridge: Cambridge University Press, 2008.

Flournoy, Théodore. *From India to the Planet Mars: A Case of Multiple Personality with Imaginary Languages*. Foreword by C. G. Jung and commentary by Mireille Cifali. Edited and introduced by Sonum Shamdasani. Princeton: Princeton University Press, 1994. Originally published as *Des Indes à la planète Mars: étude sur un cas de somnambulisme avec glossolalie* (Genève: Édition Atar, 1899), and first translated into English by Daniel B. Vermilye in 1900.

Frasca-Spada, Marina, and Nick Jardine, eds. *Books and the Sciences in History*. Cambridge: Cambridge University Press, 2000.

Frost, Edwin Brant. *An Astronomer's Life*. Boston: Houghton Mifflin, 1933.

Fyfe, Aileen. *Science and Salvation: Evangelical Popular Science Publishing in Victorian Britain*. Chicago: University of Chicago Press, 2004.

Fyfe, Aileen. "The Information Revolution." In *The Cambridge History of the Book in Britain, Volume 6, 1830–1914*, edited by David McKitterick, 567–94. Cambridge: Cambridge University Press, 2009.

Fyfe, Aileen, and Bernard Lightman, eds. *Science in the Marketplace: Nineteenth-Century Sites and Experiences*. Chicago: University of Chicago Press, 2007.

Galison, Peter. "Judgement against Objectivity." In *Picturing Science, Producing Art*, edited by Peter Galison and Caroline Jones, 327–59. New York: Routledge, 1998.

Galison, Peter. *Einstein's Clocks, Poincare's Maps: Empires of Time*. London: Hodder and Stoughton, 2003.

Garwood, Christine. *Flat Earth: The History of an Infamous Idea*. London: Pan Macmillan, 2007.

Gascoigne, S. C. B. "The Great Melbourne Telescope and Other 19th-Century Reflectors." *Historical Records of Australian Science* 10, no. 3 (1995): 223–45.

Geppert, Alexander C. T. *Fleeting Cities: Imperial Expositions in Fin-de-Siècle Europe*. Basingstoke: Palgrave Macmillan, 2010.

Gieryn, Thomas F. "Boundary Work and the Demarcation of Science from Non-Science: Strains and Interests in Professional Ideologies of Scientists." *American Sociological Review* 48 no. 6 (1983): 781–95.

Gieryn, Thomas F. *Cultural Boundaries of Science: Credibility on the Line*. Chicago: University of Chicago Press, 1999.

Gillespie, Richard. *The Great Melbourne Telescope*. Melbourne: Museum Victoria, 2011.

Gingerich, Owen. "The Satellites of Mars: Prediction and Discovery." *Journal for the History of Astronomy* 1, no. 2 (1970): 109–15.

Gingerich, Owen, ed. *The General History of Astronomy*. Vol. 4, *Astrophysics and Twentieth-Century Astronomy to 1950*. Cambridge: Cambridge University Press, 1984.

Glass, I. S. *Victorian Telescope Makers: The Lives and Letters of Thomas and Howard Grubb*. Bristol: Institute of Physics, 1997.

Goldfarb, Stephen. "Science and Democracy: A History of the Cincinnati Observatory, 1842–1872." *Ohio History* 78 (1969): 172–223.

Golinski, Jan. *The Experimental Self: Humphry Davy and the Making of a Man of Science*. Chicago: University of Chicago Press, 2016.

Gooday, Graeme. "Profit and Prophecy: Electricity in the Late-Victorian Periodical." In *Science in the Nineteenth-Century Periodical: Reading the Magazine of Nature*, by Geoffrey Cantor, Gowan Dawson, Graeme Gooday, Richard Noakes, Sally Shuttleworth, and Jonathan R. Topham, 238–54. Cambridge: Cambridge University Press, 2004.

Gooday, Graeme. "Sunspots, Weather, and the Unseen Universe: Balfour Stewart's Anti-Materialist Representations of 'Energy' in British Periodicals." In *Science Serialized: Representation of the Sciences in Nineteenth-Century Periodicals*, edited by Geoffrey Cantor and Sally Shuttleworth, 111–47. Cambridge, MA: MIT Press, 2004.

Gordin, Michael D. *Scientific Babel: How Science Was Done before and after Global English.* Chicago: University of Chicago Press, 2015.

Gould, Stephen Jay. "War of the Worldviews." In *Leonardo's Mountain of Clams and the Diet of Worms: Essays on Natural History*, 339–54. London: Jonathan Cape, 1998.

Greenhalgh, Paul. *Ephemeral Vistas: The Expositions Universelles, Great Exhibitions and World's Fairs, 1851–1939.* Manchester: Manchester University Press, 1988.

Greenwood, Frederick. *The Seven Curses of London.* London: Stanley, Rivers, 1869.

Guarneri, Julia. *Newsprint Metropolis: City Papers and the Making of Modern Americans.* Chicago: University of Chicago Press, 2017.

Guthke, Karl S. *The Last Frontier: Imagining Other Worlds, from the Copernican Revolution to Modern Science Fiction.* Translated by Helen Atkins. Ithaca, NY: Cornell University Press, 1990.

Hacking, Ian. "Telepathy: Origins of Randomization in Experimental Design." *Isis* 79, no. 3 (1988): 427–51.

Hackmann, Willem. "The Magic Lantern for Scientific Enlightenment." In *Learning by Doing: Experiments and Instruments in the History of Science*, edited by Peter Heering and Roland Wittje, 113–39. Stuttgart: Franz Steiner, 2011.

Hackmann, Willem. "Spectacular Science through the Magic Lantern." *Bulletin of the Scientific Instrument Society* 116 (Mar. 2013): 30–41.

Hampton, Mark. "Rethinking the 'New Journalism,' 1850s–1930s." *Journal of British Studies* 43, no. 2 (2004): 278–90.

Hampton, Mark. "Representing the Public Sphere: The New Journalism and Its Historians." In *Transatlantic Print Culture, 1880–1940: Emerging Media, Emerging Modernisms*, edited by Ann Ardis and Patrick Collier, 15–29. Basingstoke: Palgrave Macmillan, 2008.

Haynes, Roslynn. *H.G. Wells, Discoverer of the Future: The Influence of Science on His Thought.* London: Macmillan, 1980.

Hays, J. N. "The London Lecturing Empire, 1800–50." In *Metropolis and Province: Science in British Culture, 1780–1850*, edited by Ian Inkster and Jack Morrell, 91–119. London: Hutchinson, 1983.

Hearnshaw, John B. *The Analysis of Starlight: Two Centuries of Astronomical Spectroscopy.* Cambridge: Cambridge University Press, 2014.

Heffernan, William C. "The Singularity of our Inhabited World: William Whewell and A. R. Wallace in Dissent." *Journal of the History of Ideas* 39, no. 1 (1978): 81–100.

Hendrix, Howard V., George Edgar Slusser, and Eric S. Rabkin, eds. *Visions of Mars: Essays on the Red Planet in Fiction and Science.* Jefferson, NC: McFarland, 2011.

Henson, Louise, Geoffrey Cantor, Gowan Dawson, Richard Noakes, Sally Shuttleworth, and Jonathan R. Topham. *Culture and Science in the Nineteenth-Century Media.* Aldershot: Ashgate, 2004.

Herf, Jeffrey. *Reactionary Modernism: Technology, Culture, and Politics in Weimar and the Third Reich.* Cambridge: Cambridge University Press, 1984.

Herschel, John. *A Treatise on Astronomy.* London: Longman, Rees, Orme, Brown, Green and Longman, 1833.

Herschel, John. *Outlines of Astronomy.* London: Longman, Brown, Green, and Longmans, 1849.

Hetherington, Norriss S. "Amateurs versus Professionals: the British Association and the Controversy over the Canals on Mars." *Journal of the British Astronomical Association* 86, no. 4 (1976): 303–8.

Hetherington, Norriss S. "Percival Lowell: Professional Scientist or Interloper?" *Journal of the History of Ideas* 42, no. 1 (1981): 159–61.

Hetherington, Norriss S. "Mid-Nineteenth-Century American Astronomy: Science in a Developing Nation." *Annals of Science* 40, no. 1 (1983): 61–80.

Hetherington, Norriss S. *Science and Objectivity: Episodes in the History of Astronomy.* Ames: Iowa State University Press, 1988.

Hewitt, Martin. "Beyond Scientific Spectacle: Image and Word in Nineteenth-Century Popular Lecturing." In *Popular Exhibitions, Science and Showmanship, 1840–1910*, edited by Joe Kember, John Plunkett, and Jill A. Sullivan, 79–95. London: Pickering and Chatto, 2012.

Hewitt, Martin. *The Dawn of the Cheap Press in Victorian Britain: The End of the 'Taxes on Knowledge,' 1849–1869.* London: Bloomsbury, 2013.

Higgitt, Rebekah. "A British National Observatory: The Building of the New Physical Observatory at Greenwich, 1889–1898." *British Journal for the History of Science* 47, no. 4 (2014): 609–35.

Highham, John. "The Matrix of Specialization." In *The Organization of Knowledge in Modern America, 1860–1920*, edited by Alexandra Oleson and John Voss, 3–18. Baltimore: Johns Hopkins University Press, 1979.

Hilgartner, Stephen. "The Dominant View of Popularization: Conceptual Problems, Political Uses." *Social Studies of Science* 20, no. 3 (1990): 519–39.

Hobsbawm, Eric. *The Age of Capital, 1848–1875.* London: Abacus, 1997.

Hochadel, Oliver. "Atapuerca: The Making of a Magic Mountain, Popular Science Books and Human-Origins-Research in Contemporary Spain." In *Communicating Science in 20th-Century Europe: A Survey on Research and Comparative Perspectives*, edited by Arne Schirrmacher, 149–63. Max-Planck-Institut für Wissenschaftsgeschichte Preprint 385. Berlin: Max Planck Institute for the History of Science, 2009.

Hodgkinson, Ruth. *The Origins of the National Health Service: The Medical Services of the New Poor Law, 1834–1871.* London: Wellcome Historical Medical Library, 1967.

Hofling, Charles K. "Percival Lowell and the Canals of Mars." *British Journal of Medical Psychology* 37 (1964): 33–42.

Holden, Edward S. *Mountain Observatories in America and Europe.* Washington, DC: Smithsonian Institution, 1896.

Hollis, H. P. "The Decade 1870–1880." In *History of the Royal Astronomical Society, 1820–1920,* edited by J. L. E. Dreyer and H. H. Turner, 167–211. London: Royal Astronomical Society, 1923.

Hopkins, Eric. *A Social History of the English Working Classes.* London: Hodder and Stoughton, 1979.

Hoppen, K. Theodore. *The Mid-Victorian Generation, 1846–1886.* Oxford: Clarendon, 1998.

Hopwood, Nick, Simon Schaffer, and Jim Secord. "Seriality and Scientific Objects in the Nineteenth Century." *History of Science* 48, no. 3/4 (2010): 251–85.

Hoskin, Michael. "The First Drawing of a Spiral Nebula." *Journal for the History of Astronomy* 13, no. 2 (1982): 97–101.

Hoskin, Michael. "John Herschel's Cosmology." *Journal for the History of Astronomy* 18, no. 1 (1987): 1–34.

Hoskin, Michael. "Rosse, Robinson, and the Resolution of the Nebulae." *Journal for the History of Astronomy* 21, no. 4 (1990): 331–34.

Howsam, Leslie. "An Experiment with Science for the Nineteenth-Century Book Trade: The International Scientific Series." *British Journal for the History of Science* 33, no. 2 (2000): 187–207.

Hoyt, William G. *Lowell and Mars.* Tucson: University of Arizona Press, 1976.

Hufbauer, Karl. *Exploring the Sun: Solar Science since Galileo.* Baltimore: Johns Hopkins University Press, 1991.

Hughes, Arthur. "Science in English Encyclopaedias, 1704–1874." Part 1, *Annals of Science* 7, no. 4 (1951): 340–70; Part 2, *Annals of Science* 8, no. 4 (1952): 323–67; Part 3, *Annals of Science* 9, no. 3 (1953): 233–64; Part 4, *Annals of Science* 11, no. 1 (1955): 74–92.

Hutchins, Roger, *British University Observatories, 1772–1939.* Farnham: Ashgate, 2008.

Hyde, W. Lewis. "The Calamity of the Great Melbourne Telescope." *Proceedings of the Astronomical Society of Australia* 7, no. 2 (1987): 227–30.

Jackson, Kate. *George Newnes and the New Journalism in Britain, 1880–1910: Culture and Profit.* Aldershot: Ashgate, 2001.

Jaki, Stanley. *Planets and Planetarians: A History of Theories of the Origin of Planetary Systems.* Edinburgh: Scottish Academic Press, 1978.

James, Mary A. *Elites in Conflict: The Antebellum Clash over the Dudley Observatory.* New Brunswick, NJ: Rutgers University Press, 1987.

James, William. *The Principles of Psychology.* 2 vols. New York: Henry Holt, 1890.

James, William. *The Will to Believe and Other Essays in Popular Philosophy*. New York: Longmans, Green, 1897.

James, William. *Pragmatism: A New Name for Some Old Ways of Thinking*. New York: Longmans, Green, 1907.

James, William. *Essays in Radical Empiricism*. New York: Longman Green, 1912.

Jansen, Axel. *Alexander Dallas Bache: Building the American Nation through Science and Education in the Nineteenth Century*. Frankfurt: Campus Verlag, 2011.

Jardine, Boris. "Between the Beagle and the Barnacle: Darwin's Microscopy, 1837–1854." *Studies in History and Philosophy of Science* 40, no. 4 (2009): 382–95.

Jardine, Boris. "Made Real: Artifice and Accuracy in Nineteenth-Century Scientific Illustration." *Science Museum Group Journal* 2 (2014). http://dx.doi.org /10.15180/140208.

Jardine, Nick. "Books, Texts, and the Making of Knowledge." In *Books and the Sciences in History*, edited by Marina Frasca-Spada and Nick Jardine, 393–407. Cambridge: Cambridge University Press, 2000.

Jarvis, Adrian. *Samuel Smiles and the Construction of Victorian Values*. Stroud: Sutton, 1997.

Jensen, J. Vernon. "The X Club: Fraternity of Victorian Scientists." *British Journal for the History of Science* 5, no. 1 (1970): 63–72.

Jones, Aled. *Powers of the Press: Newspapers, Power and the Public in Nineteenth-Century England*. Aldershot: Scolar, 1996.

Jones, Bessie Z., and Lyle G. Boyd. *The Harvard College Observatory: The First Four Directorships, 1839–1919*. Cambridge, MA: Belknap Press, 1971.

Jordan, Ellen. *The Women's Movement and Women's Employment in Nineteenth Century Britain*. London: Routledge, 1999.

Joyce, Patrick. *Visions of the People: Industrial England and the Question of Class, 1848–1914*. Cambridge: Cambridge University Press, 1993.

Juergens, George. *Joseph Pulitzer and the New York World*. Princeton: Princeton University Press, 1966.

Kaalund, Nanna Katrine Lüders. "A Frosty Disagreement: John Tyndall, James David Forbes, and the Early Formation of the X-Club." *Annals of Science* 74, no. 4 (2017): 282–98.

Kaplan, Richard L. *Politics and the American Press: The Rise of Objectivity, 1865–1920*. Cambridge: Cambridge University Press, 2002.

Keene, Melanie. "Familiar Science in Nineteenth-Century Britain." *History of Science* 52, no. 1 (2014): 53–71.

Kember, Joe. *Marketing Modernity: Victorian Popular Shows and Early Cinema*. Exeter: University of Exeter Press, 2009.

Kember, Joe, John Plunkett, and Jill A. Sullivan. *Popular Exhibitions, Science and Showmanship, 1840–1910*. London: Pickering and Chatto, 2012.

Kern, Stephen, *The Culture of Time and Space, 1880–1918.* Cambridge, MA: Harvard University Press, 1983.

King, Henry C. *The History of the Telescope.* New York: Dover, 1979.

Knelman, Judith. *Twisting in the Wind: The Murderess and the English Press.* Toronto: University of Toronto Press, 1988.

Knight, David M. "Scientists and Their Publics: Popularization of Science in the Nineteenth Century." In *The Cambridge History of Science*, Vol. 5, *Modern Physical and Mathematical Sciences*, edited by Mary Jo Nye, 72–90. Cambridge: Cambridge University Press, 2003.

Kofron, John. *"Daily News."* In *Dictionary of Nineteenth-Century Journalism in Great Britain and Ireland*, edited by Laurel Brake and Marysa Demoor, 158. London: Academia Press, 2009.

Kogan, Herman. *The Great EB: The Story of the "Encyclopaedia Britannica."* Chicago: University of Chicago Press, 1958.

Koss, Stephen. *Fleet Street Radical: A. G. Gardiner and the "Daily News."* London: Allen Lane, 1973.

Koss, Stephen. *The Rise and Fall of the Political Press in Britain.* Vol. 1, *The Nineteenth Century.* London: Hamilton, 1981.

Kruse, Paul. "The Story of the *Encyclopaedia Britannica*, 1768–1943." PhD diss., University of Chicago, 1958.

LaFeber, Walter. *The Cambridge History of American Foreign Relations.* Vol. 2, *The American Search for Opportunity, 1865–1913.* Cambridge: Cambridge University Press, 1993.

Lane, K. Maria D. "Geographers of Mars: Cartographic Inscription and Exploration Narrative in Late Victorian Representations of the Red Planet." *Isis* 96, no. 4 (2005): 477–506.

Lane, K. Maria D. "Mapping the Mars Canal Mania: Cartographic Projection and the Creation of a Popular Icon." *Imago Mundi* 58, no. 2 (2006): 198–211.

Lane, K. Maria D. "Astronomers at Altitude: Mountain Geography and the Cultivation of Scientific Legitimacy." In *High Places: Cultural Geographies of Mountains, Ice and Science*, edited by Denis E. Cosgrove and Veronica Della Dora, 126–144. London: I. B. Tauris, 2009.

Lane, K. Maria D. *Geographies of Mars: Seeing and Knowing the Red Planet.* Chicago: University of Chicago Press, 2011.

Lankford, John. "Amateur versus Professional: The Transatlantic Debate over the Measurement of Jovian Longitude." *Journal of the British Astronomical Association* 89, no. 6 (1979): 574–82.

Lankford, John. "Amateurs and Astrophysics: A Neglected Aspect in the Development of a Scientific Specialty." *Social Studies of Science* 11, no. 3 (1981): 275–303.

Lankford, John. "Astronomy's Enduring Resource." *Sky and Telescope* 76 (1988): 482–83.

Lankford, John. "Amateurs versus Professionals: The Controversy over Telescope Size in Late Victorian Science." *Isis* 72, no. 1 (1981): 11–28.

Lankford, John. *American Astronomy: Community, Careers and Power, 1859–1940.* Chicago: University of Chicago Press, 1997.

Lankford, John, and Ricky L. Slavings. "The Industrialization of American Astronomy, 1880–1940." *Physics Today* 49, no. 1 (1996): 34–40.

Latour, Bruno. *Science in Action: How to Follow Scientists and Engineers through Society.* Cambridge, MA: Harvard University Press, 1987.

Laurent, John. "Science, Society and Politics in Late Nineteenth-Century England: A Further Look at Mechanics' Institutes." *Social Studies of Science* 14, no. 4 (1984): 585–619.

Lears, Jackson. *Rebirth of a Nation: The Making of Modern America, 1877–1920.* New York: Harper Perennial, 2010.

Lee, Alan J. *The Origins of the Popular Press in England.* London: Croom Helm, 1976.

Lenoir, Timothy, ed. *Inscribing Science: Scientific Texts and the Materiality of Communication.* Stanford, CA: Stanford University Press, 1998.

Lens, Sidney. *The Forging of the American Empire, from the Revolution to Vietnam: A History of US Imperialism.* London: Pluto Press, 2003.

Lenzen, Victor F. "Charles S. Peirce as Astronomer." In *Studies in the Philosophy of Charles Sanders Peirce*, 2nd ser., edited by Edward C. Moore and Richard S. Robin, 33–50. Amherst: University of Massachusetts Press, 1964.

Lequeux, James. "The Great Nineteenth Century Refractors." *Experimental Astronomy* 25, nos. 1–3 (2009): 43–61.

Levin, Miriam, Sophie Forgan, Martina Hessler, Robert H. Kargon, and Morris Low. *Urban Modernity: Cultural Innovation in the Second Industrial Revolution.* Cambridge, MA: MIT Press, 2010.

Levine, Lawrence W. *Highbrow / Lowbrow: The Emergence of Cultural Hierarchy in America.* Cambridge, MA: Harvard University Press, 1988.

Lewenstein, Bruce V. "From Fax to Facts: Communication in the Cold Fusion Saga." *Social Studies of Science* 25, no. 3 (1995): 403–36.

Lewis, Oscar. *George Davidson: Pioneer West Coast Scientist.* Berkeley: University of California Press, 1954.

Lightman, Bernard. "Astronomy for the People: R. A. Proctor and the Popularization of the Victorian Universe." In *Facets of Faith and Science*, vol. 3: *The Role of Beliefs in the Natural Sciences*, edited by Jitse M. van der Meer, 31–45. Lanham, MD: University Press of America, 1996.

Lightman, Bernard. "Constructing Victorian Heavens: Agnes Clerke and the 'New Astronomy.'" In *Natural Eloquence: Women Reinscribe Science*, edited by

Barbara T. Gates and Ann B. Shteir, 61–75. Wisconsin: University of Wisconsin Press, 1997.

Lightman, Bernard, ed. *Victorian Science in Context*. Chicago: University of Chicago Press, 1997.

Lightman, Bernard. "The Visual Theology of Victorian Popularizers of Science: From Reverent Eye to Chemical Retina." *Isis* 91, no. 4 (2000): 651–80.

Lightman, Bernard. "*Knowledge* Confronts *Nature:* Richard Proctor and Popular Science Periodicals." In *Culture and Science in the Nineteenth-Century Media*, edited by Louise Henson, 199–221. London: Ashgate, 2004.

Lightman, Bernard. "Celestial Objects for Common Readers: Webb as a Populariser of Science." In *The Stargazer of Hardwicke: The Life and Work of Thomas William Webb*, edited by Janet Robinson and Mark Robinson, 215–34. Leominster: Gracewing, 2006.

Lightman, Bernard. "Lecturing in the Spatial Economy of Science." In *Science in the Marketplace: Nineteenth-Century Sites and Experiences*, edited by Aileen Fyfe and Bernard Lightman, 97–132. Chicago: University of Chicago Press, 2007.

Lightman, Bernard. *Victorian Popularizers of Science: Designing Nature for New Audiences*. Chicago: University of Chicago Press, 2007.

Lightman, Bernard. "Science and the Public." In *Wrestling with Nature: From Omens to Science*, edited by Ronald L. Numbers, Michael H. Shank, and Peter Harrison, 337–75. Chicago: University of Chicago Press, 2011.

Lightman, Bernard. "The Creed of Science and Its Critics." In *The Victorian World*, edited Martin Hewitt, 449–65. London: Routledge, 2012.

Lightman, Bernard. "Popularizers, Participation and the Transformations of Nineteenth-Century Publishing: From the 1860s to the 1880s." *Notes and Records of the Royal Society* 70, no. 4 (2016): 343–59.

Lightman, Bernard, and Michael S. Reidy, eds. *The Age of Scientific Naturalism: Tyndall and His Contemporaries*. London: Pickering and Chatto, 2014.

Lindquist, Jason H. "'The Mightiest Instrument of the Physical Discoverer': The Visual 'Imagination' and the Victorian Observer." *Journal of Victorian Culture* 13, no. 2 (2008): 171–99.

Lockyer, Mary, and Winifred Lockyer. *The Life and Work of Sir Norman Lockyer*. London: Macmillan, 1928.

Loomis, Elias. *The Recent Progress of Astronomy, Especially in the United States*. New York: Harper and Bros., 1856.

Loveland, Jeff. "Unifying Knowledge and Dividing Disciplines: The Development of Treatises in the *Encyclopaedia Britannica*." *Book History* 9 (2006): 57–87.

Lovell, Sir Bernard, ed. *The Royal Institution Library of Science*. Vol. 1, *Astronomy*. Barking: Elsevier, 1970.

Lowell, Percival. *Mars*. Boston: Houghton, Mifflin, 1895.

Lowell, Percival. *Mars and Its Canals*. New York: Macmillan, 1906.

Lowell, Percival. *Mars as the Abode of Life*. New York: Macmillan, 1908.

Lucier, Paul. "The Professional and the Scientist in Nineteenth-Century America." *Isis* 100, no. 4 (2009): 699–732.

Luckhurst, Roger, Laurel Brake, James Mussell, and Ed King, eds. *W. T. Stead: Newspaper Revolutionary*. London: British Library, 2012.

Lurie, Edward. *Louis Agassiz: A Life in Science*. Chicago: University of Chicago Press, 1960.

Macdonald, Lee. *Kew Observatory and the Evolution of Victorian Science, 1840–1910*. Pittsburgh: University of Pittsburgh Press, 2018.

MacLeod, Roy M. "Securing the Foundations." *Nature* 224, no. 5218 (1969): 441–44.

MacLeod, Roy M. "Seeds of Competition." *Nature* 224, no. 5218 (1969): 431–34.

MacLeod, Roy M. "The X-Club: A Social Network of Science in Late-Victorian England." *Notes and Records of the Royal Society of London* 24, no. 2 (1970): 305–22.

MacLeod, Roy M. "The Support of Victorian Science: The Endowment of Research Movement in Great Britain, 1868–1900." *Minerva* 9, no. 2 (1971): 197–230.

MacLeod, Roy M. "Evolutionism, Internationalism and Commercial Enterprise in Science: The International Scientific Series, 1871–1910." In *Development of Science Publishing in Europe*, edited by A. J. Meadows, 63–93. Amsterdam: Elsevier, 1980.

Mandell, Richard D. *Paris, 1900: The Great World's Fair*. Toronto: Toronto University Press, 1967.

Markley, Robert. *Dying Planet: Mars in Science and the Imagination*. Durham, NC: Duke University Press, 2005.

Martin, Jonathan D. "'The Grandest and Most Cosmopolitan Object Teacher': Buffalo Bill's Wild West and the Politics of American Identity, 1883–1899." *Radical History Review* 66 (1996): 93–123.

Marvin, Carolyn. *When Old Technologies Were New: Thinking about Electric Communication in the Late Nineteenth Century*. New York: Oxford University Press, 1988.

Mays, Kelly J. "The Disease of Reading and Victorian Periodicals." In *Literature in the Marketplace: Nineteenth-Century British Publishing and Reading Practices*, edited by John O. Jordan and Robert L. Patten, 165–94. Cambridge: Cambridge University Press, 1995.

McCormmach, Russell. "Ormsby MacKnight Mitchel's 'Sidereal Messenger,' 1846–1848." *Proceedings of the American Philosophical Society* 110, no. 1 (1966): 35–47.

McDermott, Robert A. "Introduction." In *Essays in Psychical Research*, edited by William James, xiii–xxxvi. Cambridge, MA: Harvard University Press, 1986.

McKitterick, David. *A History of Cambridge University Press*. Vol. 3, *New Worlds for Learning, 1873–1972*. Cambridge: Cambridge University Press, 2004.

McLaughlin-Jenkins, Erin. "Common Knowledge: Science and the Late Victorian Working-Class Press." *History of Science* 39, no. 4 (2001): 445–65.

McLaughlin-Jenkins, Erin. "Walking the Low Road: The Pursuit of Scientific Knowledge in Late Victorian Working-Class Communities." *Public Understanding of Science* 12, no. 2 (2003): 147–66.

McLean, Steven. *The Early Fiction of H. G. Wells: Fantasies of Science*. Basingstoke: Palgrave Macmillan, 2009.

Meadows, A. J. *Science and Controversy: A Biography of Sir Norman Lockyer*. Cambridge, MA: MIT Press, 1972.

Menand, Louis. "An Introduction to Pragmatism." In *Pragmatism: A Reader*, xi–xxxiv. New York: Vintage, 1997.

Menand, Louis. *The Metaphysical Club*. London: Flamingo, 2002.

Miller, Howard S. *Dollars for Research: Science and Its Patrons in Nineteenth-Century America*. Seattle: University of Washington Press, 1970.

Miller, Lillian B., Frederick Voss, and Jeannette M. Hussey. *The Lazzaroni: Science and Scientists in Mid-Nineteenth-Century America*. Washington, DC: Smithsonian Institution Press, 1972.

Miller, R. Kalley. *The Romance of Astronomy*. London: Macmillan, 1875.

Mitchel, Ormsby MacKnight. *Popular Astronomy: A Concise Elementary Treatise on the Sun, Planets, Satellites and Comets*. New York: Phinney, Blakeman and Mason, 1860.

Molvig, Ole. "Cosmological Revolutions: Relativity, Astronomy, and the Shaping of a Modern Universe." PhD diss., Princeton University, 2006.

Morgan, Charles. *The House of Macmillan (1843–1943)*. London: Macmillan, 1943.

Morrell, Jack, and Arnold Thackray. *Gentlemen of Science: Early Years of the British Association for the Advancement of Science*. Oxford: Clarendon, 1981.

Morse, Edward S. *Mars and Its Mystery*. Boston: Little, Brown, 1906.

Morus, Iwan R. "Different Experimental Lives: Michael Faraday and William Sturgeon." *History of Science* 30, no. 1 (1992): 1–28.

Morus, Iwan R. "Manufacturing Nature: Science, Technology and Victorian Consumer Culture." *British Journal for the History of Science* 29, no. 4 (1996): 403–34.

Morus, Iwan R. *When Physics Became King*. Chicago: University of Chicago Press, 2005.

Morus, Iwan R. "Seeing and Believing Science." *Isis* 97, no. 1 (2006): 101–10.

Morus, Iwan R. "'More the Aspect of Magic than Anything Natural': The Philosophy of Demonstration." In *Science in the Marketplace: Nineteenth-Century Sites and Experiences*, edited by Aileen Fyfe and Bernard Lightman, 336–70. Chicago: University of Chicago Press, 2007.

Morus, Iwan R. "Worlds of Wonder: Sensation and the Victorian Scientific Performance." *Isis* 101, no. 4 (2010): 806–16.

Morus, Iwan R., Simon Schaffer, and James A. Secord. "Scientific London." In *London: World City, 1800–1840*, edited by Celina Fox, 129–42. New Haven, CT: Yale University Press, 1992.

Moyer, Albert E. *A Scientist's Voice in American Culture: Simon Newcomb and the Rhetoric of Scientific Method.* Berkeley: University of California Press, 1992.

Mullen, Richard D. "The Undisciplined Imagination: Edgar Rice Burroughs and Lowellian Mars." In *SF: The Other Side of Realism, Essays on Modern Fantasy and Science Fiction*, edited by Thomas D. Clareson, 229–47. Bowling Green, OH: Bowling Green University Popular Press, 1971.

Mussell, James. *Science, Time, and Space in the Late Nineteenth-Century Periodical Press: Movable Types.* Aldershot: Ashgate, 2007.

Mussell, James. "Arthur Cowper Ranyard, *Knowledge* and the Reproduction of Astronomical Photographs in the Late Nineteenth-Century Periodical Press." *British Journal for the History of Science* 42, no. 3 (2009): 345–80.

Mussell, James. "'Characters of Blood and Flame': Stead and the Tabloid Campaign." In *W. T. Stead: Newspaper Revolutionary*, edited by Roger Luckhurst, Laurel Brake, James Mussell, and Ed King, 22–36. London: British Library, 2012.

Nall, Joshua. Review of *Geographies of Mars: Seeing and Knowing the Red Planet*, by K. Maria D. Lane. *British Journal for the History of Science* 45, no. 4 (2012): 692–94.

Nall, Joshua. "Constructing Canals on Mars: Event Astronomy and the Transmission of International Telegraphic News." *Isis* 108, no. 2 (2017): 280–306.

Nasim, Omar W. *Observing by Hand: Sketching the Nebulae in the Nineteenth Century.* Chicago: University of Chicago Press, 2013.

Nasim, Omar W. "James Nasmyth on the Moon: or, On Becoming a Lunar Being, without the Lunacy." In *Selene's Two Faces: From 17th Century Drawings to Spacecraft Imaging*, edited by Carmen Pérez González, 147–87. Leiden: Brill, 2018.

Nasmyth, James, and James Carpenter. *The Moon: Considered as a Planet, a World, and a Satellite.* London: John Murray, 1874.

Nead, Lynda. *The Haunted Gallery: Painting, Photography, Film c. 1900.* New Haven, CT: Yale University Press, 2008.

Nerone, John, and Kevin Barnhurst. "Stead in America." In *W. T. Stead: Newspaper Revolutionary*, edited by Roger Luckhurst, Laurel Brake, James Mussell, and Ed King, 98–114. London: British Library, 2012.

Newcomb, Simon. *Popular Astronomy.* New York: Harper and Bros., 1878.

Newcomb, Simon. *His Wisdom the Defender.* New York: Harper and Bros., 1900.

Newcomb, Simon. *Astronomy for Everybody: A Popular Exposition of the Wonders of the Heavens.* New York: McClure, Phillips, 1902.

Newcomb, Simon. *The Reminiscences of an Astronomer.* New York: Harper and Bros., 1903.

Nichol, John Pringle. *Views of the Architecture of the Heavens: In a Series of Letters to a Lady.* Edinburgh: William Tate, 1837.

Nicholson, Tony. "The Provincial Stead." In *W. T. Stead: Newspaper Revolutionary*, edited by Roger Luckhurst, Laurel Brake, James Mussell, and Ed King, 7–21. London: British Library, 2012.

Nisbett, Catherine E. "Business Practice: The Rise of American Astrophysics, 1859–1919." PhD diss., Princeton University, 2007.

Nisbett Becker, Catherine. "Professionals on the Peak." *Science in Context* 22, no. 3 (2009): 487–507.

Noakes, Richard. "'Cranks and Visionaries': Science, Spiritualism and Transgression in Victorian Britain." PhD diss., University of Cambridge, 1998.

Noakes, Richard. "Spiritualism, Science and the Supernatural in Mid-Victorian Britain." In *The Victorian Supernatural*, edited by Nicola Bown, Carolyn Burdett, and Pamela Thurschwell, 23–43. Cambridge: Cambridge University Press, 2004.

Noakes, Richard. "The 'World of the Infinitely Little': Connecting Physical and Psychical Realities Circa 1900." *Studies in History and Philosophy of Science* 39, no. 3 (2008): 323–34.

Norberg, Arthur L. "Simon Newcomb's Early Astronomical Career." *Isis* 69, no. 2 (1978): 209–25.

Norberg, Arthur L. "Simon Newcomb's Role in the Astronomical Revolution of the Early Nineteen Hundreds." In *Sky with Ocean Joined: Proceedings of the Sesquicentennial Symposia of the US Naval Observatory, December 5–8, 1980*, edited by Steven J. Dick and LeRoy E. Doggett, 74–88. Washington, DC: U.S. Naval Observatory, 1983.

Nye, David E. "Electrifying Expositions, 1880–1939." In *Fair Representations: World's Fairs and the Modern World*, edited by Robert W. Rydell, Nancy E. Gwinn, and James Gilbert, 140–56. Amsterdam: VU University Press, 1994.

Nye, Mary Jo. *Before Big Science: The Pursuit of Modern Chemistry and Physics, 1800–1940.* New York: Twayne, 1996.

O'Connor, Richard. *The Scandalous Mr. Bennett.* Garden City, NY: Doubleday, 1962.

O'Connor, Ralph. *The Earth on Show: Fossils and the Poetics of Popular Science, 1802–1856.* Chicago: University of Chicago Press, 2008.

O'Connor, Ralph. "Reflections on Popular Science in Britain: Genres, Categories, and Historians." *Isis* 100, no. 2 (2009): 333–45.

Olson, Richard G. "The Gould Controversy at Dudley Observatory: Public and Professional Values in Conflict." *Annals of Science* 27, no. 3 (1971): 265–76.

Oppenheim, Janet. *The Other World: Spiritualism and Psychical Research in England, 1850–1914.* Cambridge: Cambridge University Press, 1988.

Oreskes, Naomi. "The Rejection of Continental Drift." *Historical Studies in the Physical Sciences* 18, no. 2 (1988): 311–48.

Örnebring, Henrik. "The Maiden Tribute and the Naming of Monsters: Two Case Studies of Tabloid Journalism as Alternative Public Sphere." *Journalism Studies* 7, no. 6 (2006): 851–68.

Osterbrock, Donald E. *James E. Keeler, Pioneer American Astrophysicist: And the Early Development of American Astrophysics.* Cambridge: Cambridge University Press, 1984.

Osterbrock, Donald E. "The Rise and Fall of Edward S. Holden." Part 1, *Journal for the History of Astronomy* 15, no. 2 (1984): 81–127; Part 2, *Journal for the History of Astronomy* 15, no. 3 (1984): 151–76.

Osterbrock, Donald E. "To Climb the Highest Mountain: W. W. Campbell's 1909 Mars Expedition to Mount Whitney." *Journal for the History of Astronomy* 20, no. 2 (1989): 77–97.

Osterbrock, Donald E. *Yerkes Observatory, 1892–1950: The Birth, Near Death, and Resurrection of a Scientific Research Institution.* Chicago: University of Chicago Press, 1997.

Osterbrock, Donald E., John R. Gustafson, and W. J. Shiloh Unruh. *Eye on the Sky: Lick Observatory's First Century.* Berkeley: University of California Press, 1988.

Otis, Laura. *Networking: Communicating with Bodies and Machines in the Nineteenth Century.* Ann Arbor: University of Michigan Press, 2001.

Otis, Laura. "The Metaphoric Circuit: Organic and Technological Communication in the Nineteenth Century." *Journal of the History of Ideas* 63, no. 1 (2002): 105–28.

Pandora, Katherine. "Popular Science in National and Transnational Perspective: Suggestions from the American Context." *Isis* 100, no. 2 (2009): 346–58.

Pang, Alex Soojung-Kim. "The Social Event of the Season: Solar Eclipse Expeditions and Victorian Culture." *Isis* 84, no. 2 (1993): 252–77.

Pang, Alex Soojung-Kim. "Victorian Observing Practices, Printing Technology, and Representations of the Solar Corona, (1): The 1860s and 1870s." *Journal for the History of Astronomy* 25, no. 4 (1994): 249–74.

Pang, Alex Soojung-Kim. "Victorian Observing Practices, Printing Technology, and Representations of the Solar Corona, (2): The Age of Photomechanical Reproduction." *Journal for the History of Astronomy* 26, no. 1 (1995): 63–75.

Pang, Alex Soojung-Kim. "'Stars Should Henceforth Register Themselves': Astrophotography at the Early Lick Observatory." *British Journal for the History of Science* 30, no. 2 (1997): 177–202.

Pang, Alex Soojung-Kim. "Technology, Aesthetics, and the Development of Astro-photography at the Lick Observatory." In *Inscribing Science: Scientific Texts and the Materiality of Communication*, edited by Timothy Lenoir, 223–48. Stanford, CA: Stanford University Press, 1998.

Pang, Alex Soojung-Kim. *Empire and the Sun: Victorian Solar Eclipse Expeditions.* Stanford, CA: Stanford University Press, 2002.

Pettitt, Clare. *Dr. Livingstone, I Presume? Missionaries, Journalists, Explorers and Empire.* London: Profile, 2007.

Pickering, William H. *A Photographic Atlas of the Moon.* Cambridge, MA: Harvard College Observatory, 1903.

Pickering, William H. *Mars.* Boston: Richard G. Badger, 1921.

Pillinger, Colin. *Beagle: From Sailing Ship to Mars Spacecraft.* Milton Keynes: XNP Productions, 2003.

Plotkin, Howard. "Edward C. Pickering and the Endowment of Scientific Research in America, 1877–1918." *Isis* 69, no. 1 (1978): 44–57.

Plotkin, Howard. "Edward C. Pickering, the Henry Draper Memorial, and the Beginnings of Astrophysics in America." *Annals of Science* 35, no. 4 (1978): 365–77.

Plotkin, Howard. "Harvard College Observatory's Boyden Station in Peru: Origin and Formative Years, 1879–1898." In *Mundialización de la Ciencia y Cultura Nacional: Actas del Congreso Internacional "Ciencia, Descubrimiento y Mundo Colonial,"* edited by A. Lafuente, A. Elena, and M. L. Ortega, 689–705. Madrid: Doce Calles, 1993.

Plotkin, Howard. "William H. Pickering in Jamaica: The Founding of Woodlawn and Studies of Mars." *Journal for the History of Astronomy* 24, no. 1/2 (1993): 101–22.

Poovey, Mary. "The Limits of the Universal Knowledge Project: British India and the East Indiamen." *Critical Inquiry* 31 (2004): 183–202.

Porter, Roy. "Gentlemen and Geology: The Emergence of a Scientific Career, 1660–1920." *The Historical Journal* 21, no. 4 (1978): 809–36.

Porter, Theodore M. "The Fate of Scientific Naturalism: From Public Sphere to Professional Exclusivity." In *Victorian Scientific Naturalism: Community, Identity, Continuity*, edited by Gowan Dawson and Bernard Lightman, 265–87. Chicago: University of Chicago Press, 2014.

Powell, Kerry. "The Saturday Review." In *British Literary Magazines, Part 3: The Victorian and Edwardian Age, 1837–1913*, edited by Alvin Sullivan, 379–83. Westport, CT: Greenwood Press, 1984.

Procter, Ben. *William Randolph Hearst: The Early Years, 1863–1910.* Oxford: Oxford University Press, 1998.

Proctor, Richard A. *Remarks on Browning's Stereograms of Mars.* London: John Browning, 1869.

Proctor, Richard A. *Other Worlds Than Ours: The Plurality of Worlds Studied under the Light of Recent Scientific Researches*. London: Longmans, Green, 1870.

Proctor, Richard A. *Light Science for Leisure Hours*. London: Longmans, Green, 1871.

Proctor, Richard A. *Essays on Astronomy*. London: Longmans, Green, 1872.

Proctor, Richard A. *The Borderland of Science*. London: Smith, Elder, 1873.

Proctor, Richard A. *The Moon: Her Motions, Aspect, Scenery and Physical Condition*. Manchester: Alfred Brothers, 1873.

Proctor, Richard A. *The Transits of Venus: A Popular Account of Past and Coming Transits*. London: Longmans, Green, 1874.

Proctor, Richard A. *The Universe and the Coming Transits*. London: Longmans, Green, 1874.

Proctor, Richard A. *Science Byways*. London: Smith, Elder, 1875.

Proctor, Richard A. *Wages and Wants of Science-Workers*. London: Smith, Elder, 1876.

Proctor, Richard A. *The Life and Death of a World*. R. A. Proctor's Lectures, no. 1. Sydney: Carmichael and Co, 1880.

Proctor, Richard A. *Rough Ways Made Smooth: A Series of Familiar Essays on Scientific Subjects*. New York: R. Worthington, 1880.

Proctor, Richard A. *The Poetry of Astronomy*. London: Smith, Elder, 1881.

Proctor, Richard A. *Other Suns Than Ours*. London, W. H. Allen, 1887.

Proctor, Richard A. *Watched by the Dead: A Loving Study of Dickens' Half-Told Tale*. London: W. H. Allen, 1887.

Proctor, Richard A. *Old and New Astronomy*. Completed by A. Cowper Ranyard. London: Longman, Green, 1892.

Putnam, William L. *The Explorers of Mars Hill: A Centennial History of Lowell Observatory, 1894–1994*. West Kennebunk, ME: Phoenix, 1994.

Putnis, Peter, Chandrika Kaul, and Jürgen Wilke, eds. *International Communication and Global News Networks: Historical Perspectives*. New York: Hampton, 2011.

Radway, Janice A. *A Feeling for Books: The Book-of-the-Month Club, Literary Taste, and Middle-Class Desire*. Chapel Hill: University of North Carolina Press, 1997.

Rankin, Jeremiah, and Ruth Barton. "Tyndall, Lewes and Popular Representations of Scientific Authority in Victorian Britain." In *The Age of Scientific Naturalism: Tyndall and His Contemporaries*, edited by Bernard Lightman and Michael S. Reidy, 51–70. London: Pickering and Chatto, 2014.

Ratcliff, Jessica. *The Transit of Venus Enterprise in Victorian Britain*. London: Pickering and Chatto, 2008.

Ray, Angela G. *The Lyceum and Public Culture in the Nineteenth-Century United States*. East Lansing: Michigan State University Press, 2005.

Reingold, Nathan. *Science in Nineteenth Century America: A Documentary History.* London: Macmillan, 1966.

Reingold, Nathan. "Alexander Dallas Bache: Science and Technology in the American Idiom." *Technology and Culture* 11, no. 2 (1970): 163–77.

Reingold, Nathan. "Definitions and Speculations: The Professionalization of Science in America in the Nineteenth Century." In *The Pursuit of Knowledge in the Early American Republic: American Scientific and Learned Societies from Colonial Times to the Civil War*, edited by Alexandra Oleson and Sanborn C. Brown, 33–69. Baltimore: Johns Hopkins University Press, 1976.

Reingold, Nathan. *Science, American Style.* New Brunswick, NJ: Rutgers University Press, 1991.

Richards, Evelleen. "Redrawing the Boundaries: Darwinian Science and Victorian Women Intellectuals." In *Victorian Science in Context*, edited by Bernard Lightman, 119–42. Chicago: University of Chicago Press, 1997.

Richardson, Robert D. *William James: In the Maelstrom of American Modernism.* Boston: Houghton Mifflin Harcourt, 2006.

Riffenburgh, Beau. *The Myth of the Explorer: The Press, Sensationalism and Geographical Discovery.* London: Belhaven Press, 1993.

Robbins, William G. *Colony and Empire: The Capitalist Transformation of the American West.* Lawrence: University Press of Kansas, 1994.

Robertson, Frances. "Science and Fiction: James Nasmyth's Photographic Images of the Moon." *Victorian Studies* 48, no. 4 (2006): 595–623.

Robinson, W. Sydney. *Muckraker: The Scandalous Life and Times of W. T. Stead.* London: Robson Press, 2012.

Romein, Jan. *The Watershed of Two Eras: Europe in 1900.* Middletown, CT: Wesleyan University Press, 1978.

Rose, Jonathan. *The Intellectual Life of the British Working Classes.* New Haven, CT: Yale University Press, 2002.

Rossiter, Margaret W. "Benjamin Silliman and the Lowell Institute: The Popularization of Science in Nineteenth-Century America." *New England Quarterly* 44, no. 4 (1971): 602–26.

Rothenberg, Marc. "Organization and Control: Professionals and Amateurs in American Astronomy, 1899–1918." *Social Studies of Science* 11, no. 3 (1981): 305–25.

Royal Society, Southern Telescope Committee. *Correspondence concerning the Great Melbourne Telescope.* London: Taylor and Francis, 1871.

Rubin, Joan S. *The Making of Middlebrow Culture.* Chapel Hill: University of North Carolina Press, 1992.

Ruiz-Castell, Pedro. "Priority Claims and Public Disputes in Astronomy: E. M. Antoniadi, J. Comas i Solà and the Search for Authority and Social Prestige in the

Early Twentieth Century." *British Journal for the History of Science* 44, no. 4 (2011): 509–31.

Russell, Colin A. *Science and Social Change, 1700–1900.* [London]: Macmillan, 1983.

Rydell, Robert W. *All the World's a Fair: Visions of Empire at the American International Expositions, 1876–1916.* Chicago: University of Chicago Press, 1984.

Rydell, Robert W., and Rob Kroes. *Buffalo Bill in Bologna: The Americanization of the World, 1869–1922.* Chicago: University of Chicago Press, 2005.

Saarloos, Léjon. "Virtues of Courage and Virtues of Restraint: Tyndall, Tait, and the Use of the Imagination in Late Victorian Science." In *Epistemic Virtues in the Sciences and the Humanities*, edited by Jeroen van Dongen and Herman Paul, 109–28. Cham: Springer, 2017.

Sagan, Carl. *Cosmos.* New York: Random House, 1980.

Sagan, Carl, and Paul Fox. "The Canals of Mars: An Assessment after Mariner 9," *Icarus* 25 (1975): 602–12.

Salmon, Richard. "'A Simulacrum of Power': Intimacy and Abstraction in the Rhetoric of the New Journalism." In *Nineteenth-Century Media and the Construction of Identities*, edited by Laurel Brake, Bill Bell, and David Finkelstein, 27–39. Basingstoke: Palgrave, 2000.

Saum, Lewis O. "The Proctor Interlude in St. Joseph and in America: Astronomy, Romance and Tragedy." *American Studies International* 37, no. 1 (1999): 34–54.

Schaffer, Simon. "Astronomers Mark Time: Discipline and the Personal Equation." *Science in Context* 2, no. 1 (1988): 115–45.

Schaffer, Simon. "The Nebular Hypothesis and the Science of Progress." In *History, Humanity and Evolution: Essays in Honour of John C. Greene*, edited by J. R. Moore, 131–64. Cambridge: Cambridge University Press, 1989.

Schaffer, Simon. "Where Experiments End: Tabletop Trials in Victorian Astronomy." In *Scientific Practice: Theories and Stories of Doing Physics*, edited by Jed Z. Buchwald, 257–99. Chicago: University of Chicago Press, 1995.

Schaffer, Simon. "The Leviathan of Parsonstown: Literary Technology and Scientific Representation." In *Inscribing Science: Scientific Texts and the Materiality of Communication*, edited by Timothy Lenoir, 182–222. Stanford, CA: Stanford University Press, 1998.

Schaffer, Simon. "On Astronomical Drawing." In *Picturing Science, Producing Art*, edited by Caroline A. Jones and Peter Galison, 441–74. London: Routledge, 1998.

Schaffer, Simon. "Time Machines." In *The Whipple Museum of the History of Science: Instruments and Interpretations, to Celebrate the Sixtieth Anniversary of R. S. Whipple's Gift to the University of Cambridge*, edited by Liba Taub and Frances Willmoth, 345–66. Cambridge: Whipple Museum of the History of Science, 2006.

Schaffer, Simon. "A World Elsewhere." Lecture 4, Tarner Lectures, Trinity College, Cambridge, 2010. http://sms.cam.ac.uk/media/746999.

Schaffer, Simon. "Easily Cracked: Scientific Instruments in States of Disrepair." *Isis* 102, no. 4 (2011): 706–17.

Schalck, Harry G. "Fleet Street in the 1880s: The New Journalism." In *Papers for the Millions: The New Journalism in Britain, 1850s to 1914*, edited by Joel H. Wiener, 73–87. New York: Greenwood Press, 1988.

Schiller, Dan. *Objectivity and the News: The Public and the Rise of Commercial Journalism.* Philadelphia: University of Pennsylvania Press, 1981.

Schivelbusch, Wolfgang. *The Railway Journey: The Industrialization of Time and Space in the 19th Century.* Berkeley: University of California Press, 1986.

Schroeder, David A. "A Message from Mars: Astronomy and Late-Victorian Culture." PhD diss., Indiana University, 2002.

Schudson, Michael. *Discovering the News: A Social History of American Newspapers.* New York: Basic Books, 1978.

Schudson, Michael. "News, Public, Nation." *American Historical Review* 107, no. 2 (2002): 481–95.

Schults, Raymond L. *Crusader in Babylon: W. T. Stead and the Pall Mall Gazette.* Lincoln: University of Nebraska Press, 1972.

Schwarzlose, Richard A. *The Nation's Newsbrokers.* 2 vols. Evanston, IL: Northwestern University Press, 1989–90.

Sconce, Jeffrey. *Haunted Media: Electronic Presence from Telegraphy to Television.* Durham, NC: Duke University Press, 2000.

Scott, Donald M. "The Popular Lecture and the Creation of a Public in Mid-Nineteenth-Century America." *Journal of American History* 66, no. 4 (1980): 791–809.

Secord, Anne. "Corresponding Interests: Artisans and Gentlemen in Nineteenth-Century Natural History." *British Journal for the History of Science* 27, no. 4 (1994): 383–408.

Secord, Anne. "Science in the Pub: Artisan Botanists in Early Nineteenth-Century Lancashire." *History of Science* 32, no. 3 (1994): 269–315.

Secord, James A. "Progress in Print." In *Books and the Sciences in History*, edited by Marina Frasca-Spada and Nick Jardine, 369–89. Cambridge: Cambridge University Press, 2000.

Secord, James A. *Victorian Sensation: The Extraordinary Publication, Reception, and Secret Authorship of "Vestiges of the Natural History of Creation."* Chicago: University of Chicago Press, 2000.

Secord, James A. "Knowledge in Transit." *Isis* 95, no. 4 (2004): 654–72.

Secord, James A. "The Electronic Harvest." *British Journal for the History of Science* 38, no. 4 (2005): 463–67.

Secord, James A. "A Planet in Print: Rethinking the Discovery of Neptune." Unpublished paper, STS Workshop, Department of the History and Philosophy of Science, University of Cambridge, 2007.

Secord, James A. "Science, Technology and Mathematics." In *The Cambridge History of the Book in Britain*, Vol. 6, *1830–1914*, edited by David McKitterick, 443–74. Cambridge: Cambridge University Press, 2009.

Secord, James A. "Life on the Moon, Newspapers on Earth." Distinguished Lecture, History of Science Society Annual Meeting, Boston, 2013.

Secord, James A. *Visions of Science: Books and Readers at the Dawn of the Victorian Age*. Oxford: Oxford University Press, 2014.

Secord, James A. "Paper Comets: Prophecy, Panic and Public Judgement in the Nineteenth Century." Paper presented at the Annual Meeting of the History of Science Society, Toronto, Ontario, 2017.

Sedgwick, Ellery. "*The Atlantic Monthly.*" In *American Literary Magazines: The Eighteenth and Nineteenth Centuries*, edited by Edward E. Chielens, 50–57. New York: Greenwood Press, 1986.

Shapin, Steven. "The Politics of Observation: Cerebral Anatomy and Social Interests in the Edinburgh Phrenology Disputes." In *On the Margins of Science: The Social Construction of Rejected Knowledge*, edited by Roy Wallace, 139–78. Keele: University of Keele, 1979.

Shapin, Steven. "Nibbling at the Teats of Science: Edinburgh and the Diffusion of Science in the 1830s." In *Metropolis and Province: Science in British Culture, 1780–1850*, edited by Ian Inkster and Jack Morrell, 151–78. Philadelphia: University of Pennsylvania Press, 1983.

Shapin, Steven. "Science and the Public." In *Companion to the History of Modern Science*, edited by Robert C. Olby, Geoffrey N. Cantor, John R. R. Christie, and M. Jonathan S. Hodge, 990–1007. London: Routledge, 1990.

Shapin, Steven. "The Image of the Man of Science." In *The Cambridge History of Science*, Vol. 4, *Eighteenth-Century Science*, edited by Roy Porter, 159–83. Cambridge: Cambridge University Press, 2003.

Shapin, Steven, and Simon Schaffer. *Leviathan and the Air-Pump: Hobbes, Boyle, and the Experimental Life*. Princeton: Princeton University Press, 1985.

Sheehan, William. *Planets and Perception: Telescopic Views and Interpretations, 1609–1909*. Tucson: University of Arizona Press, 1988.

Sheehan, William. *The Immortal Fire Within: The Life and Work of Edward Emerson Barnard*. Cambridge: Cambridge University Press, 1995.

Sheehan, William. *The Planet Mars: A History of Observation and Discovery*. Tucson: University of Arizona Press, 1996.

Shklovskii, Iosef, and Carl Sagan. *Intelligent Life in the Universe*. San Francisco: Holden-Day, 1966.

Shoemaker, Philip S. "Stellar Impact: Ormsby MacKnight Mitchel and Astronomy in Antebellum America." PhD diss., University of Wisconsin–Madison, 1991.

Simon, Lynda. *Genuine Reality: A Life of William James*. New York: Harcourt Brace, 1998.

Skrupskelis, Ignas K., and Elizabeth M. Berkeley, eds. *The Correspondence of William James*. 12 vols. Charlottesville: University of Virginia Press, 1992–2004.

Slater, Michael, and John Drew. "Introduction." In *Dickens' Journalism*, Vol. 4, *The Uncommercial Traveller and Other Papers, 1859–70*, edited by Michael Slater and John Drew, xi–xxiii. London: Dent, 2000.

Slotkin, Richard. *The Fatal Environment: The Myth of the Frontier in the Age of Industrialization, 1800–1890*. Norman: University of Oklahoma Press, 1998.

Smith, David C., ed. *The Correspondence of H. G. Wells*. 4 vols. London: Pickering and Chatto, 1998.

Smith, Jonathan. *Charles Darwin and Victorian Visual Culture*. Cambridge: Cambridge University Press, 2009.

Smith, Michael L. *Pacific Visions: California Scientists and the Environment, 1850–1915*. New Haven, CT: Yale University Press, 1987.

Smith, Robert W. "The Cambridge Network in Action: The Discovery of Neptune." *Isis* 80, no. 3 (1989): 395–422.

Smith, Robert W. "A National Observatory Transformed: Greenwich in the Nineteenth Century." *Journal for the History of Astronomy* 22, no. 1 (1991): 5–20.

Smith, Robert W. "Martians and Other Aliens." *Studies in History and Philosophy of Biological and Biomedical Sciences* 30, no. 2 (1999): 237–54.

Smith, Crosbie, and M. Norton Wise. *Energy and Empire: A Biographical Study of Lord Kelvin*. Cambridge: Cambridge University Press, 1989.

Sobel, Dava. *The Planets*. London: Harper Perennial, 2006.

Sobel, Dava. *The Glass Universe: The Hidden History of the Women Who Took the Measure of the Stars*. London: 4th Estate, 2016.

Solnit, Rebecca. *River of Shadows: Eadweard Muybridge and the Technological Wild West*. New York: Viking, 2003.

Solomon, Matthew. "A Trip to the Fair; or, Moon-Walking in Space." In *Fantastic Voyages of the Cinematic Imagination: Georges Méliès's "Trip to the Moon,"* edited by Matthew Solomon, 143–60. Albany: State University of New York Press, 2011.

Sommer, Andreas. "Psychical Research and the Origins of American Psychology: Hugo Munsterberg, William James and Eusapia Palladino." *History of the Human Sciences* 25, no. 2 (2012): 23–44.

Sponsel, Alistair. "Constructing a 'Revolution in Science': The Campaign to Promote a Favourable Reception for the 1919 Solar Eclipse Experiments." *British Journal for the History of Science* 35, no. 4 (2002): 439–67.

Sponsel, Alistair. *Darwin's Evolving Identity: Adventure, Ambition, and the Sin of Speculation.* Chicago: University of Chicago Press, 2018.

Stachurski, Richard. *Longitude by Wire: Finding North America.* Columbia: University of South Carolina Press, 2009.

Staley, Richard. "On the Co-Creation of Classical and Modern Physics." *Isis* 96 (2005): 530–58.

Staley, Richard. "Conspiracies of Proof and Diversity of Judgement in Astronomy and Physics: On Physicists' Attempts to Time Light's Wings and Solve Astronomy's Noblest Problem." In *L'Événement Astronomique du Siècle? Histoire Sociale des Passages de Vénus, 1874–1882,* edited by David Aubin, 83–98. Nantes: Centre François Viète, 2007.

Staley, Richard. *Einstein's Generation: The Origins of the Relativity Revolution.* Chicago: University of Chicago Press, 2008.

Staley, Richard. "Michelson and the Observatory: Physics and the Astronomical Community in Late Nineteenth-Century America." In *The Heavens on Earth: Observatories and Astronomy in Nineteenth-Century Science and Culture,* edited by David Aubin, Charlotte Bigg, and H. Otto Sibum, 225–52. Durham, NC: Duke University Press, 2010.

Standage, Tom. *The Victorian Internet: The Remarkable Story of the Telegraph and the Nineteenth Century's On-Line Pioneers.* New York: Berkley, 1999.

Stanley, Henry M. *How I Found Livingstone: Travels, Adventures, and Discoveries in Central Africa.* London: Sampson Low, Marston, Low, and Searle, 1872.

Stead, W. T. *The Americanization of the World, or The Trend of the Twentieth Century.* New York: Horace Markley, 1901.

Stephens, Carlene E. "'The Most Reliable Time': William Bond, the New England Railroads, and Time Awareness in 19th-Century America." *Technology and Culture* 30, no. 1 (1989): 1–24.

Stephens, Carlene E. "Astronomy as Public Utility: The Bond Years at the Harvard College Observatory." *Journal for the History of Astronomy* 21, no. 1 (1990): 21–36.

Stevens, John D. *Sensationalism and the New York Press.* New York: Columbia University Press, 1991.

Strauss, David. "'Fireflies Flashing in Unison': Percival Lowell, Edward Morse and the Birth of Planetology." *Journal for the History of Astronomy* 24, no. 3 (1993): 157–69.

Strauss, David. "Percival Lowell, W. H. Pickering and the Founding of the Lowell Observatory." *Annals of Science* 51, no. 1 (1994): 37–58.

Strauss, David. *Percival Lowell: The Culture and Science of a Boston Brahmin.* Cambridge, MA: Harvard University Press, 2001.

Tatarewicz, Joseph N. *Space Technology and Planetary Astronomy.* Bloomington: Indiana University Press, 1990.

Tattersdill, Will. *Science, Fiction, and the* Fin-de-Siècle *Periodical Press*. Cambridge: Cambridge University Press, 2016.

Taub, Liba. *Science Writing in Greco-Roman Antiquity*. Cambridge: Cambridge University Press, 2017.

Terpak, Frances. "Imaging the Moon." In *Devices of Wonder: From the World in a Box to Images on a Screen*, edited by Barbara Maria Stafford and Frances Terpak, 197–204. Los Angeles: Getty Publications, 2001.

Thompson, E. P. "The Political Education of Henry Mayhew." *Victorian Studies* 11, no. 1 (1967): 41–62.

Thompson, Robert L. *Wiring a Continent: The History of the Telegraph Industry in the United States, 1832–1866*. Princeton: Princeton University Press, 1947.

Thurs, Daniel P. *Science Talk: Changing Notions of Science in American Popular Culture*. New Brunswick, NJ: Rutgers University Press, 2007.

Tilley, Elizabeth. "Christianity, Journalism, and Popular Print: W. T. Stead and the Salvation Army." In *W. T. Stead: Newspaper Revolutionary*, edited by Roger Luckhurst, Laurel Brake, James Mussell, and Ed King, 59–76. London: British Library, 2012.

Topham, Jonathan R. "BJHS Special Section: Book History and the Sciences—Introduction." *British Journal for the History of Science* 33, no. 2 (2000): 155–58.

Topham, Jonathan R. "Scientific Publishing and the Reading of Science in Nineteenth-Century Britain: A Historiographical Survey and Guide to Sources." *Studies in History and Philosophy of Science* 31, no. 4 (2000): 559–612.

Topham, Jonathan R. "Introduction. Focus: Historicizing 'Popular Science.'" *Isis* 100, no. 2 (2009): 310–18.

Topham, Jonathan R. "Rethinking the History of Science Popularization / Popular Science." In *Popularizing Science and Technology in the European Periphery, 1800–2000*, edited by Faidra Papanelopoulou, Agustí Nieto-Galan, and Enrique Perdiguero, 1–20. Aldershot: Ashgate, 2009.

Tucher, Andie. *Froth and Scum: Truth, Beauty, Goodness, and the Ax Murderer in America's First Mass Medium*. Chapel Hill: University of North Carolina Press, 1994.

Tucker, Jennifer. *Nature Exposed: Photography as Eyewitness in Victorian Science*. Baltimore: Johns Hopkins University Press, 2005.

Turner, Frank M. "The Victorian Conflict between Science and Religion: A Professional Dimension." *Isis* 69, no. 3 (1978): 356–76.

Turner, Frederick J. *The Significance of the Frontier in American History*. London: Penguin, 2008. Essays first published in 1893, 1896, 1909, and 1910. Page references are to the 2008 edition.

Tyndall, John. *Six Lectures on Light*. London: Longmans, Green, 1873.

Van Helden, Albert. "Telescope Building, 1850–1900." In *The General History of Astronomy*, Vol. 4, *Astrophysics and Twentieth-Century Astronomy to 1950*, edited by Owen Gingerich, 40–58. Cambridge: Cambridge University Press, 1984.

Vetter, Jeremy. *Field Life: Science in the American West During the Railroad Era.* Pittsburgh: University of Pittsburgh Press, 2016.

Via Galveston. [New York]: Central and South American Telegraph Company, [1895].

Wallace, Alfred Russel. *The Wonderful Century: The Age of New Ideas in Science and Invention.* 2nd ed. London: Swan Sonnenschein, 1903.

Wallace, Alfred Russel. *Is Mars Habitable? A Critical Examination of Percival Lowell's Book "Mars and Its Canals," with an Alternative Explanation.* London: Macmillan, 1907.

Walsh, S. Padraig. *Anglo-American General Encyclopedias: A Historical Bibliography, 1703–1967.* New York: R. R. Bowker, 1968.

Warner, Deborah J. *Alvan Clark and Sons: Artists in Optics.* Washington, DC: Smithsonian Institution Press, 1968.

Warner, Deborah J. "Astronomy in Antebellum America." In *The Sciences in the American Context: New Perspectives,* edited by Nathan Reingold, 55–75. Washington, DC: Smithsonian Institution Press, 1979.

Wayman, Dorothy. *Edward Sylvester Morse: A Biography.* Cambridge, MA: Harvard University Press, 1942.

Webb, George E. "The Planet Mars and Science in Victorian America." *Journal of American Culture* 3, no. 4 (1980): 573–80.

Webb, George E. *Tree Rings and Telescopes: The Scientific Career of A. E. Douglass.* Tucson: University of Arizona Press, 1983.

Wells, H. G. *The Time Machine.* London: William Heinemann, 1895. Reprinted with introduction by Marina Warner and notes by Steven McLean. Edited by Patrick Parrinder. London: Penguin, 2005.

Wells, H. G. *The War of the Worlds.* London: William Heinemann, 1898.

White, Paul. *Thomas Huxley: Making the "Man of Science."* Cambridge: Cambridge University Press, 2003.

White, Paul. "The Conduct of Belief: Agnosticism, The Metaphysical Society, and the Formation of Intellectual Communities." In *Victorian Scientific Naturalism: Community, Identity, Continuity,* edited by Gowan Dawson and Bernard Lightman, 220–41. Chicago: University of Chicago Press, 2014.

White, Richard. *"It's Your Misfortune and None of My Own": A New History of the American West.* Norman: University of Oklahoma Press, 1991.

Whyte, Frederic. *The Life of W. T. Stead.* 2 vols. London: J. Cape, 1925.

Wiener, Joel H. "How New Was the New Journalism?" In *Papers for the Millions: The New Journalism in Britain, 1850s to 1914,* 47–71. New York: Greenwood Press, 1988.

Wiener, Joel H. "Introduction." In *Papers for the Millions: The New Journalism in Britain, 1850s to 1914*, xi–xix. New York: Greenwood Press, 1988.

Wiener, Joel H., ed. *Papers for the Millions: The New Journalism in Britain, 1850s to 1914.* New York: Greenwood Press, 1988.

Wiener, Joel H. "The Americanization of the British Press, 1830–1914." In *Studies in Newspaper and Periodical History, 1994 Annual*, edited by Michael Harris and Tom O'Malley, 61–74. Westport, CT: Greenwood Press, 1996.

Wiener, Joel H. *The Americanization of the British Press, 1830s–1914: Speed in the Age of Transatlantic Journalism.* Basingstoke: Macmillan, 2011.

Wiener, Joel H., and Mark Hampton, eds. *Anglo-American Media Interactions, 1850–2000.* Basingstoke: Macmillan, 2007.

Williams, Mari E. W. "Astronomy in London: 1860–1900." *Quarterly Journal of the Royal Astronomical Society* 28, no. 1 (1987): 10–26.

Williams, Raymond. *The Long Revolution.* Harmondsworth: Penguin Books, 1965.

Willis, Artemis. "'What the Moon Is Like': Technology, Modernity, and Experience in a Late-Nineteenth-Century Astronomical Entertainment." *Early Popular Visual Culture* 15, no. 2 (2017): 175–203.

Willis, Martin. *Mesmerists, Monsters, and Machines: Science Fiction and the Cultures of Science in the Nineteenth Century.* Kent, OH: Kent State University Press, 2006.

Willis, Martin. *Vision, Science and Literature, 1870–1920: Ocular Horizons.* London: Pickering and Chatto, 2011.

Willis, Martin, ed. *Staging Science: Scientific Performance on Street, Stage and Screen.* London: Palgrave Macmillan, 2016.

Winseck, Dwayne R., and Robert M. Pike. *Communication and Empire: Media, Markets, and Globalization, 1860–1930.* Durham, NC: Duke University Press, 2007.

Winter, Alison. "The Construction of Orthodoxies and Heterodoxies in the Early Victorian Life Sciences." In *Victorian Science in Context*, edited by Bernard Lightman, 24–50. Chicago: University of Chicago Press, 1997.

Winter, Alison. *Mesmerized: Powers of Mind in Victorian Britain.* Chicago: University of Chicago Press, 1998.

Winter, Frank H. "The 'Trip to the Moon' and Other Early Spaceflight Simulation Shows, ca. 1901–1915: Part I." In *History of Rocketry and Astronautics: Proceedings of the Twenty–Eighth and Twenty-Ninth History Symposia of the International Academy of Astronautics, Jerusalem, Israel, 1994, Oslo, Norway, 1995*, edited by Donald C. Elder and Christophe Rothmund, 133–61. San Diego, CA: American Astronautical Society, 2001.

Winter, Frank H. "The 'Trip to the Moon' and Other Early Spaceflight Simulation Shows, ca. 1901–1915: Part II." In *History of Rocketry and Astronautics:*

Proceedings of the Thirtieth History Symposia of the International Academy of Astronautics: Jerusalem, Beijing, China, 1996, edited by Hervé Moulin and Donald C. Elder, 3–28. San Diego, CA: American Astronautical Society, 2003.

Worth, Aaron. "Imperial Transmissions: H. G. Wells, 1897–1901." *Victorian Studies* 53, no. 1 (2010): 65–89.

Wright, Helen. *Explorer of the Universe: A Biography of George Ellery Hale.* New York: Dutton, 1966.

Wright, Helen. *James Lick's Monument: The Saga of Captain Richard Floyd and the Building of the Lick Observatory.* Cambridge: Cambridge University Press, 1987.

Yeo, Richard R. "Reading Encyclopedias: Science and the Organization of Knowledge in British Dictionaries of Arts and Sciences, 1730–1850." *Isis* 82, no. 1 (1991): 24–49.

Yeo, Richard R. "Encyclopaedic Knowledge." In *Books and the Sciences in History*, edited by Marina Frasca-Spada and Nick Jardine, 207–24. Cambridge: Cambridge University Press, 2000.

Yeo, Richard R. *Encyclopaedic Visions: Scientific Dictionaries and Enlightenment Culture.* Cambridge: Cambridge University Press, 2001.

Zochert, Donald. "Science and the Common Man in Ante-Bellum America." *Isis* 65, no. 4 (1974): 448–73.

SIMON NEWCOMB'S ENCYCLOPEDIA WORK

The following are encyclopedias carrying articles attributable to Simon Newcomb, listed in chronological order. Full lists of the specific articles by Newcomb are given in Archibald, "Simon Newcomb, 1835–1909." Further details on the encyclopedias themselves can be found in Walsh, *Anglo-American General Encyclopedias.*

Encyclopaedia Britannica. 9th ed. 24 vols. Edinburgh: Adam and Charles Black, 1875–88.

"Moon."

Johnson's New Universal Cyclopaedia. 1st ed. 4 vols. New York: A. J. Johnson's and Son, 1876–78.

"Astronomy" and three other articles.

Johnson's Universal Cyclopaedia. 3rd ed. 8 vols. New York: A. J. Johnson, 1893–97.

"Astronomy" and seventy-one other articles. Newcomb was associate editor for astronomy and mathematics.

New American Supplement to the Encyclopedia Britannica [unauthorized]. 5 vols. New York: Werner Company, 1897.

"Astronomy" and two other articles.

The American Educator: A Library of Universal Knowledge. 6 vols. Philadelphia: Syndicate Publishing Company, 1897.

Newcomb's involvement with this publication is unclear. He is listed as one of the "associate editors and special contributors," but specific articles are unsigned and are not listed in Archibald, "Simon Newcomb, 1835–1909." In "How an Encyclopaedia May Be Edited," *Nation* 87 (Nov. 19, 1908), 492, Newcomb repudiates any connection with the *Twentieth Century Encyclopaedia* (1901) but admits that "it is quite possible that it contains one or two articles which I wrote some ten or twenty years ago for a cyclopaedia that has since become extinct." Archibald (66) implies that the *American Educator* is the original work in question. It is notable that at least three distinct reprints / pirate copies of this work entered circulation: *The Universal Cyclopaedia and Dictionary* (Chicago: National Book Concern, 1898); *The Universal Reference Library* (New York and Chicago: Dictionary and Cyclopedia Company, 1900); *Twentieth Century Encyclopaedia: A Library of Universal Knowledge* (Philadelphia: Syndicate Publishing Company, 1901).

Universal Cyclopaedia. 4th ed. 12 vols. New York: D. Appleton, 1900.

"Astronomy" and seventy other articles. Newcomb was associate editor for astronomy and mathematics.

Encyclopaedia Britannica. 10th ed. Eleven supplementary volumes added to 9th ed. Edinburgh: Adam and Charles Black, 1902–3.

"Astronomy" and four other articles.

Encyclopedia Americana. 2nd ed. 16 vols. New York and Chicago: R. S. Peale, 1902.

"Astronomy" and twenty-one other articles. Newcomb was listed as an associate and advisory editor.

Encyclopaedia Britannica. 11th ed. 29 vols. Cambridge: Cambridge University Press, 1910–11.

"Astronomy, Descriptive" and twenty-one other articles, including "Mars." Newcomb was associate editor for astronomy.

INDEX

Note: Page numbers in *italics* refer to figures.

astrophysics (*cont.*): and professionalization, 13–15, 26–28, 73–76,
79–80, 192n25 (*see also* professionalization, historiographic
models of); and technological
West, 87–92; and telegraphic news
distribution, 3–4, 80–82, 84–86,
107–33; U.S.A., rapid growth of in,
71–72, 78. *See also* new astronomy;
planetary science; spectroscope;
telescopes
Atlantic Monthly, 76, 83, 85, 139

Bache, Alexander Dallas, 73, 209n14
Bailey, Solon, 117–18, 166, *167*
Ball, Robert, 70
Barker, George, 111
Barnard, E. E., 136, 150–51, 157, 165,
166, 169, 233n98
Bennett, James Gordon, Jr., 52, 108–9,
112
Bennett, James Gordon, Sr., 108–9
Besant, Annie, 51
Bessel, Friedrich, 73
Bly, Nellie, 83
books, circulation figures for, 225n24
Borderland, 102
Borderland of Science, The (Proctor),
45
Boston Commonwealth, 135
Boston Daily Globe, 122
Boston Journal, 4
Boston Scientific Society, 137
Boyden Fund, 109, 114, 117
Boyden Station, 109–10, 114–23, 130,
213n51
Boyden, Uriah Atherton, 86, 109,
213n51
Brera Observatory, 9
Brewster, David, 33, 191n13
British Association for the Advance-
ment of Science (BAAS), 43,
58–59
British Astronomical Association, 154,
174, 224n11, 213n75
British Journal of Medical Psychology,
176
Brontë, Charlotte, 65

Brooklyn Times, 124
Browning, John, 34, 62, 195n24
Brun, Ingeborg, *11*
Buffalo Bill, 90
Bunsen, Robert, 5, 24
Busk, Hans, the Younger, *11*
Butler, Josephine, 51

cable telegraphy. *See* telegraphy
Cambridge University Press, 148
Campbell, W. W., 124, 137, 151, 174,
224n13, 227n43
canali. *See under* canals on Mars
canals on Mars: 1892 opposition,
observed during, 124–29, 132, 137;
1909 opposition, not observed by
major observatories during, 174–5;
artificial explanation for, 10–12, *11*,
22, 125, 138–42, 151, 170; "canali,"
translation of, 128–29, 142, 150,
177, 191n17, 222n67; decline of
theory of, 175, 181; discovery by
Schiaparelli, 9–10, 21, 94; experience in observing, importance of,
151–54, 158–59, 165–71, 228n53;
gemination of (i.e. seen double),
10, 124–25, 132, 151; Mariner
photographs, lack of correspondence with, 177, 235n10; natural
explanations for, 10, 115, 126, 127,
174; as news commodity, 124–29,
132; as optical illusion, 150–54,
157, 161–62, 165–71, *167–68*, 177;
photographs of, 157–58, 159, 162,
170, 173, 230n66, 230n72, 233n96,
234n104; popular sensation, as
product of, 10–12, 70, 105–6, 137,
142, 175, 177; *Times* (London),
reported in, 21–24. *See also* Mars:
mapping of; Lowell, Percival;
Schiaparelli, Giovanni
Carnegie, Andrew, 51
Carpenter, James, 62, *64*
Carr, John W., 160
Catalogue of Scientific Papers (Royal
Society), 37, 38–39
Cattell, James McKeen, 160, 210n22
Cayley, Arthur, 58–59, 195n25

Henry Draper Observatory, 82
heroism, 87–90, 108
Herschel, John, 41–43, 46
Herschel, William, 41
His Wisdom the Defender (Newcomb), 145
Hofling, Charles K., 176
Holden, Edward: conservative approach, criticized for, 123–24, 131; *Encyclopaedia Britannica*, author for, 150; Flammarion, debate with over life on Mars, 116, 123; Lick Observatory, director of, 78; Lowell, criticism of, 137; on Martian canal observations at Lick in 1892, 124–25, 127, 132; mountain observatories, advocate for, 89; *Mountain Observatories in America and Europe*, 89; Newcomb, understudy to, 77–78; newspapers, criticism of as forums for astronomical debate, 116, 119, 127; Proctor, conflict with over *Atlantic Monthly* review, 76–78, 85; Proctor, conflict with over Lick telescope, 83–86, 89, 146–47; and technological West, 87
Hooper, Horace, 148
Huggins, William, 9, 24, 35, 82, 84
Hussey, William, 124
Huxley, Thomas Henry, 77, 100, 197n33

imagination, use of in the sciences, 35, 40–46, 60–66, 70, 75–76, 94–102, 138
imaginative astronomy: decline of, 69–79, 92–93, 103; definition of, 35; at expositions, 155, *156*; Flammarion's work as, 206n135; Lowell influenced by, 136–42; Newcomb as critic of, 45–46, 75–77, 210n22; new journalism and, 46–67; Proctor's work as, 35–46
imperialism, American, 112
inhabited Mars hypothesis. *See* Mars: life on
International Scientific Series, 65
interplanetary communication. *See* signaling

Jackson, William, 148
James, William, 162–65, 169, 232n86. *See also* pragmatism
Janssen, Jules, 30
Javelle, Stéphane, 98, 99, 100, 101
journalism. *See* new journalism; newspapers
Journal of the British Astronomical Association, 96
journals, circulation figures for, 16, 39–40, 199n59
Jupiter, 45, 60, 76, 95–96

Kant, Immanuel, 33
Kelvin, Lord, 143–44
Kirchhoff, Gustav, 5, 24
Knowledge, 56–57, 198n54, 199n57

Lampland, Carl Otto, 230n72
Lancet, 48
Land Nationalisation (Wallace), 65
Lane, K. Maria D., 10, 33, 70, 88
Lazzaroni, 73–74, 80
lectures and lecturing: attendance figures, 16, 53, 65, 75, 203n103, 204n109; fees for, 204n109, 207n140; and imaginative astronomy, 44–46, 65, 138; Lowell and, 135–37, 173; lyceum lecture system, 54; and "platform culture," 65; Proctor and, 23, 25–26, 45, 53–55, 60, 65, 69, 75, 203n103, 206n131, 207n140; social stratification of, 55, 204n107; Sunday Lecture Society, 55
Ledger, Edmund, 158
Lick, James, 83, 87, 89
Lick Observatory: Holden appointed director of, 78; Proctor critical of its telescope 83–86; site and seeing conditions, criticism of, 122, 151; site and seeing conditions, praise of, 92–93, 95, 109, 150, 220n51; technological West, exemplary site of, 87–90; William Pickering, as site of opposition to, 118, 122–124, 131, 132
life on Mars. *See* Mars: life on

Porter, Theodore, 36
pragmatism, 162–65, 169–70. *See also* canals on Mars: as optical illusion; James, William; Peirce, Charles
press syndication. *See under* telegraphy
Pritchard, Charles, 30–31, 37–40
Proceedings of the Royal Society, 139
Proctor, Richard Anthony: Airy, conflict with, 28–31, 44; Americanization of British science and society, advocate of, 52–55; analogy of Mars to Earth, 9, 33–35, 61, 62; astronomer, credentials as, 25–26, 30, 33, 39, 58; Astronomer Royal, as candidate to become, 51; author on astronomy, as widely-read, 21–23, 25–26, 28–29, 53, 66; *The Borderland of Science*, 45; canals on Mars, letter to London *Times* concerning, 21–24, 46, 61, 66, 69–70; directorship of Draper Observatory, attempt to secure, 82; early career of, 25–28, 33; encyclopedias, author and editor for, 146–47, 149; *English Mechanic*, contributions to, 39–40; and globes of Mars, *11*, 33–34, 62; Holden, conflict with, 76–78, 83–86, 89, 146–47; on imagination in the sciences, 35, 40–46, 60–66; on John Herschel as ideal astronomer, 41–43; *Knowledge*, editor of, 56–57; large telescopes, critic of, 83–86, 92; as lecturer, 23, 25–26, 45, 53–55, 60, 65, 69, 75, 203n103, 206n131, 207n140; Lockyer, conflict with, 27–31, 37–40, 195n22; mapping of Mars, 9, *11*, 33, *34*, 62; Newcomb, conflict with, 75–79; and new journalism, 50–67; *Other Worlds than Ours*, 9, *34*, 34–40; personal life of, 25, 69, 93; planetary evolution, theory of, 60–61, 66, 92, 138, 223n7; on pluralism (plurality of worlds), 33–35, 60–61, 92; *The Poetry of Astronomy*, 44; "popularizer," characterized as, 22–23, 25, 30, 38,

40, 58, 76, 85, 119, 177; positional astronomy, critic of, 27, 77; RAS Gold Medal, nomination for, 30, 51; RAS *Monthly Notices*, editor of, 25, 29, 31; scientific naturalists, opponent of, 27–28, 65–66; on star-drift and structure of the universe, 25–26, 30, 33, 42; state funding of science, critic of, 43; and Stead, 50–54, 57–59, 66; on transit of Venus enterprise, 28–29; on visual perception in astronomical observation, 152; on visual versus mathematical representation, 45–46, 58–65; *Wages and Wants of Science-Workers*, 43, 77
professional astronomy. *See* professionalization, historiographic models of
professionalization, historiographic models of, 12, 13–15, 66, 73–76, 179–81, 191n21, 192n25
projections on Mars. *See* signaling
psychical research, 143, 163–64
psychology of perception. *See* canals on Mars: as optical illusion; Newcomb, Simon: visual perception, experiments on
Publications of the Astronomical Society of the Pacific, 137
Pulitzer, Joseph, 56, 83–84
Punch, 95–96
Putnam's Monthly, 54

Reader, 32
Reform Act of 1867, 48
Reingold, Nathan, 74
relativity, 181–82
Respighi, Lorenzo, 30
Review of Reviews, 102, 202n92, 226n34
Rhodes, Cecil, 52
Ritchie, John, 82
Rowland, Henry, 78, 79, 209n11
Royal Astronomical Society (RAS), 25–31, 33, 37, 44, 46, 70, 79
Royal Institution, 25, 44, 55, 197n35, 204n107

Royal Observatory, 26, 29, 31, 41, 70–71, 118–19
Royal Polytechnic Institution, 55
Royal Society, 37, 38–39
Rutherfurd, Lewis M., 62, *63*

Sagan, Carl, 177, 235n6, 235n8, 235n10
St. Louis Globe-Democrat, 211n41, 213n48
Saint Louis Post-Dispatch, 122
sales figures. *See* circulation figures for
Salvation Army, 51
San Francisco Chronicle, 122, 123, 212n46, 219n28
San Francisco Daily Examiner, 84–85, 89–90, 212n46
Saturday Review, 36, 51–52, 101
Saturn and Its System (Proctor), 33
Saunder, Samuel Arthur, 174
Schaeberle, John Martin, 124
Schaffer, Simon, 71
Schiaparelli, Giovanni: canals on Mars, 1877 mapping of, 9–10, *11*, 127; canals on Mars, reception of his observations, 21–22, 70, 94, 116, 124–26, 129, 132, 142, 150, 191n17, 208n10, 222n67; *Encyclopaedia Britannica* 11th edition, potential author for "Mars" article, 159; eyesight of, 153; life on Mars, views on, 174, 221n54
science fiction, 13, 101–3, 139, 175
science, funding of, 29–30, 43
scientific Lazzaroni. *See* Lazzaroni
scientific naturalism, 27–28, 31, 32, 36–40, 44, 60, 66
scientific romance. *See* science fiction
scoop (news reporting), 122–23
Scrymser, James, 112, 121
Secord, James, 106
seeing, astronomical: atmospheric conditions for, 72, 89, 92–93, 99, 109, 119, 120, 151, 153, 214n63; and event astronomy, 107–9; experimental testing of, 153; and observations of Mars, 72, 93–94, 99, 117, 120, 122, 136, 151; and

technological West, 89–90, 92–93, 103, 178–79, 214n63
Self-Help (Smiles), 48
Shinn, Milicent, 131
Sidereal Messenger (f. 1846, Cincinnati, OH), 80, 213n49
Sidereal Messenger (f. 1882, Northfield, MN), 85, 213n49
signaling: from Earth to Mars, 83, 94–103, 117, 182–83; from Mars to Earth, 94–103, 117, 182–83
Slipher, Vesto, 170
Smiles, Samuel, 48
Smithsonian Institution, 81
Social Problems (George), 65
Society for the Protection of Children, 51
solar physics. *See* astrophysics; new astronomy
Solnit, Rebecca, 87, 93, 145
solar eclipse. *See under* expeditions
South America, 112–14, 118. *See also* Boyden Station
Spectator (London), 28
spectroscope, 5, 9, 24–25, 34–36, 60, 71, 84. *See also* astrophysics; new astronomy
Spence, Edward, 86
spiritualism. *See under* Mars
Stanford, Leland, 87
Stanley, Henry Morton, 109
Stead, William Thomas, 50–58, 66, 102, 148, 226n34
stellar astronomy, 6, 71–72, 109, 130
Strange, Alexander, 29–30
Sullivan, Edmund J., *97*
Sunday Lecture Society (SLS), 55
Swift, Jonathan, 100
syndication. *See under* telegraphy

Tait, P. G., 44
taxes on knowledge, 47, 56
technological West: and the annihilation of time and space, 87–88, 93; Boyden Station as exemplary of, 114; definition of, 87; and event astronomy, 107–8;

Wesley, W. H., 157
West, American. *See* technological
   West
Whewell, William, 33, 191n13
Wiggins, E. Stone, 85, 123
Williams, Mattieu, 59
Willis, Martin, 94
Winnecke, Friedrich August Theodor, 77
Wolfe, Henry, 112–13
women's rights, 48, 57
world's fairs. *See* expositions
Wundt, Wilhelm, 163, 165

X Club, 27, 32, 36, 46, 65–66, 74

"Yankee Circus on Mars, A," 157
yellow journalism, 7, 52, 54, 56, 83. *See
also* new journalism
Yeo, Richard, 147, 149
Yerkes Observatory, 90, *91*, 213n51,
   214n63